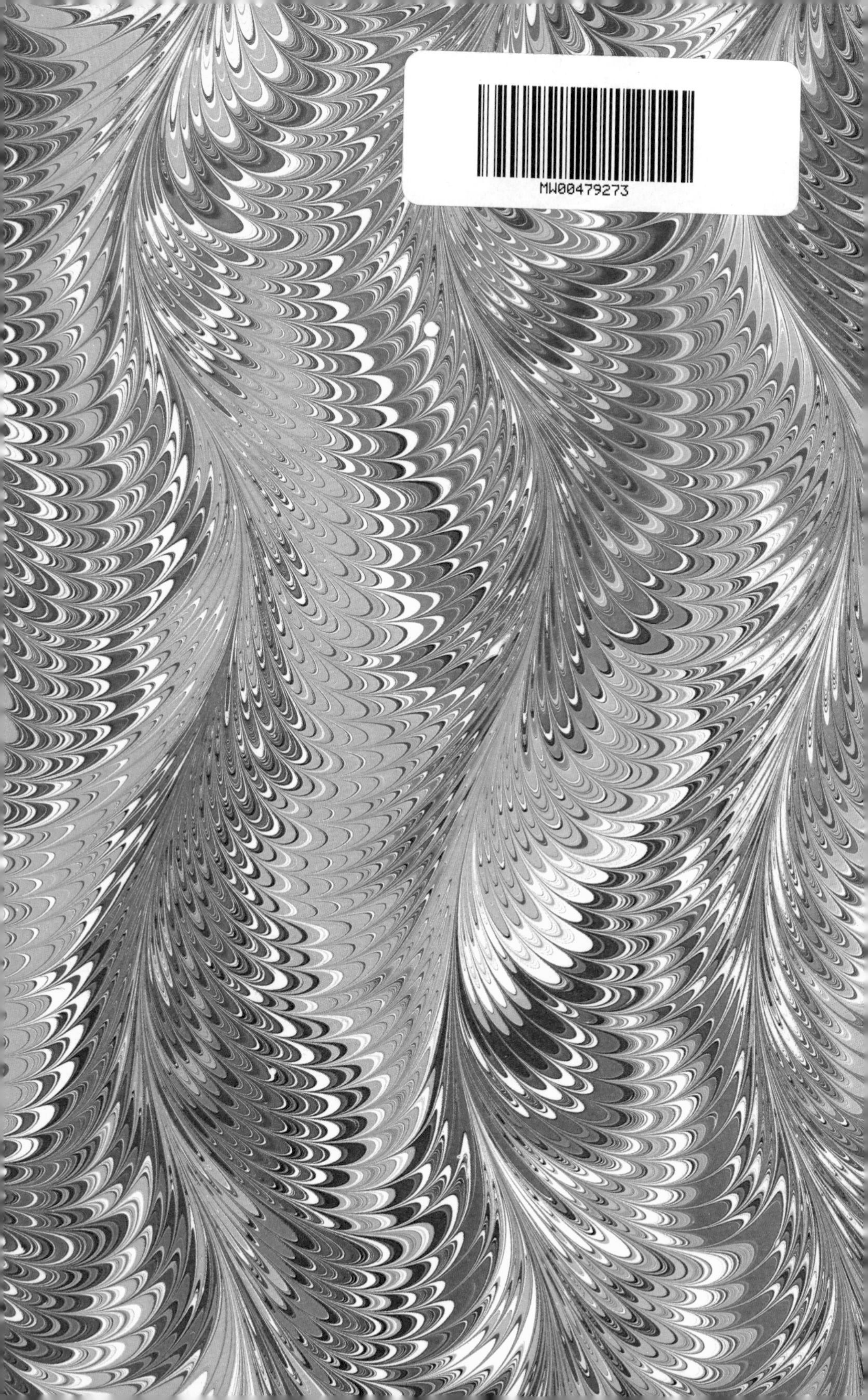
MW00479273

Flying Tiger
to
Air Commando

Chuck Baisden
33rd Pursuit Sqdn
8th Pursuit Group
1939-1941

Flying Tiger to Air Commando

Charles Baisden

Schiffer Military History
Atglen, PA

ACKNOWLEDGMENTS

To my wife, Willa, for her encouragement and typing skills in making an improved 2nd edition possible.

To my son, Dan, for his English comprehension and knowledge in the editing of this edition.

Book Design by Ian Robertson.

Library of Congress Catalog Number: 98-86280

Printed in China
ISBN: 0-7643-0690-1

Published by Schiffer Publishing Ltd.
4880 Lower Valley Road
Atglen, PA 19310
Phone: (610) 593-1777
FAX: (610) 593-2002
E-mail: Schifferbk@aol.com.
Visit our web site at: www.schifferbooks.com
Please write for a free catalog.
This book may be purchased from the publisher.
Please include $3.95 postage.
Try your bookstore first.

In Europe, Schiffer books are distributed by:
Bushwood Books
6 Marksbury Road
Kew Gardens
Surrey TW9 4JF
England
Phone: 44 (0)181 392-8585
FAX: 44 (0)181 392-9876
E-mail: Bushwd@aol.com.

Try your bookstore first.

Contents

FOREWORD

A dedicated enlisted airman recounts his experiences of an extraordinary and varied military career.

One of the best of the AVG Flying Tiger armorers.

Then an aircrew gunner in B25 bombers in another unique but unheralded Air Commando unit in Burma during which he was decorated for exemplary performance.

And even more combat duty as an aircrew gunner on B-29s in the Korean war with follow-up duty as an aircrew boom operator in tankers.

This is a book that should inspire young airmen, and I would encourage them to read it.

Major General Charles R. Bond, Jr., U.S.AF (Ret.)
(Formerly Vice Squad Leader,
lst Pursuit Squadron, AVG Flying Tigers)

BIOGRAPHY

Although this is a biography of my enlisted career in the Military, perhaps it would be of interest to the reader if I gave as background some incidents from my childhood, and the chronological order of my career.

I was born in Scranton, Pennsylvania, in 1920, and while I attended schools in the city, summers were spent in the country with my grandparents. This is where I was happiest, enjoying a life of hunting, fishing, and trapping. At the age of 14, my father gave me my first .22 rifle. At 16 I received my father's 12 gauge L.C. Smith shotgun. To pay for my hunting license and ammunition I sold muskrat pelts, and soon rabbits, grouse, and squirrel went into the cookpot.

Until I was a junior in high school, I had the dream of pursuing a career as a forester. However, after my father informed me there was no money for college, I gave up that notion. These were the days of the Great Depression of the late 1920s, and jobs were scarce and money tight. While we never went hungry, there was certainly no money for extras—like college.

While in high school, I joined the local National Guard as a rifleman. This was Company B of the 109th Infantry Regiment. We were armed with bolt-action Springfield rifles, and the folks in the community called us "dollar dummies" because our pay was $1.00 per drill night.

After graduation from high school I began working as a machinist's helper in a nylon manufacturing company, and later for my father in his jewelry manufacturing shop. One day I passed the post office and saw a poster urging high school graduates to join the U.S. Army Air Corps. Sounded good to me, and in 1939 at the age of nineteen I enlisted as a private.

In 1941 I volunteered for service in the American Volunteer Group and was the first volunteer to reach the AVG base in Burma in July of that year. During that year

I drove the Burma Road via truck. I returned to the U.S. in September of 1942, and reenlisted in the Air Force, returning to the China-Burma theater in 1943 and flying fifty-eight combat missions in B-25s as turret gunner with the First Air Commando Group. During the Korean War I served as an armorer in a fighter squadron, then flew as a gunner in B-29s from Okinawa. I finished my career in the Air Force as a KC97 Tanker boom operator, refueling SAC B-47 and B-52 bombers.

I graduated from the NCO Academy, and my awards included the Distinguished Flying Cross, the Bronze Star, the Air Medal, 4 Unit Presidential Unit Citations, and the China National Service Medal.

I am now retired and live with my wife, Willa, in Savannah, Georgia. We have two sons, a daughter, and eight grandchildren.

Master Sergeant Charles "Chuck" Baisden—studio portrait taken in Karachi, India - January 1944.

INTRODUCTION

I did not start out to write a book. Instead, I began cleaning out my papers, records, and photos that I have accumulated over many years. In doing so, I began to remember an incredible time in my life. It was a period, I realized, in which I had a chance to do some things very few people ever get to do.

As an enlisted man, I did not make policy, but I did make policies happen. I knew the military needed me to make the system work, no matter how small a part I was asked to play. And I played my part to the best of my ability for my own survival and for those who depended on me.

I cannot begin to cover the bonds of companionship with comrades who shared this part of my life. Those who have been in similar situations will know what I mean.

As to this period of my life, I have no regrets.

Charles (Chuck) Baisden, M/Sgt U.S.AF (Ret.)

1

The AVG

I enlisted in the U.S. Air Corps in 1939 as a private. My base pay was $21.00 per month. After completing a short basic training program, I was assigned to the 33rd Pursuit Squadron, 8th Pursuit Group at Langley Field, Virginia. Afterward, I was sent to Aircraft Armament School at Lowry Field, Colorado. There we learned the function and repair of .30 and .50 caliber Browning machine guns, gun sight harmonization or bore sighting, bomb shackles, pyrotechnics, synchronization, and some chemical warfare. It was only a three-month course, and thus by itself did not make me a qualified armament specialist, but it did give me the basics. The practical applications I learned by actually working in the armament shop and on the flight line.

I remember several incidents from those early years in the Army Air Corps. First, it was not an easy life. Some behavior was simply not tolerated. If a guy was drunk, didn't like to shower, or resorted to thievery, he was dealt with in no uncertain terms. He found it was better to take a shower at regular times rather than to get a cold shower with lots of soap applied with a scrub brush by people who had to live with him.

Familiarity between officers and enlisted men was not tolerated. Pilots, however, were prone to be a bit friendly. Enlisted men had to march in formation everywhere while on duty, and our cotton coveralls had to be kept fully buttoned with no sleeves rolled up.

We pulled kitchen police duty by the week and night hangar guard duty about once every two months. Night hangar duty meant you were locked in the hangar, armed with a .45 pistol, a flashlight, and an old steel cot to sleep on. I never got used to the strange sounds coming from the hangar floor where the aircraft were

Bell Airacuda YFM-1, Langley Field, VA 1940. Assigned to 8th Pursuit Group.

Curtiss P36A, Langley Field, VA 1940. Assigned to 33rd Pursuit Squadron, 8th Pursuit Group.

parked. Sleep was difficult because all the noises made it seem as if the aircraft were talking to each other within the darkened void of that huge building.

In that same building, late one night while I was on duty, an overly eager officer attempted to crawl through a window above my cot. He was about halfway through the opening when he heard the slide of my .45 come back. He rather hastily identified himself, then he chewed me out for leaving the window open. When I told him I needed the air for ventilation, I also said, "Sir, you wouldn't have made it another ten feet." He seemed shaken up about the whole thing, and I never heard about it again.

I learned the hard way never to lend one of my tools to anybody when a wrench I had loaned out was found jamming the joy stick of a P-36. Another lesson I learned was the importance of having a field telephone. It happened one day when I was on the wing of a P-40 firing the .50 caliber machine gun. We had not been able to obtain field phones for the test, but the firing range had an officer and a NCO who were acting as range safeties. Just as I was about to press a screw driver on the firing mechanism, I looked up and saw one of the range detail standing on a ladder fixing the target. My legs turned to water, and I slid off the wing to the ground below. The safety NCO asked me what happened. All I could do was point down range. We both realized I had come within a fraction of a second of killing a man. Although I had been given the order to fire, the range safeties had not seen a man who had gone back to the target to cover a bullet hole with a target paster. If I had fired the next round, it would have hit him directly in the head. We had no trouble securing a field telephone after this incident.

One year after enlisting, in the summer of 1940, I checked out as a tow reel operator on the Martin B-10 bomber. I can't count the hours I spent towing targets over Fort Monroe and Virginia Beach so the Coast Artillery and our own P-36s and P-40s could fire on them.

In November of 1940, the 8th moved to Mitchel Field, Long Island, New York. The Air Corps expanded so that one squadron of the 8th (the 33rd) was made cadre for a new group, the 57th. I was assigned to the 65th Pursuit Squadron, commanded by Captain Phil Cochran, who was later represented in Milt Caniff's cartoon, "Steve Canyon," as Flip Corkin and General Philerie.

In the spring of 1941, our entire squadron was approached by Skip Adair while we were on gunnery exercises at a small field near Windsor Locks, Connecticut. Mr. Adair was recruiting volunteers to go to China to form a P-40 Pursuit group that would protect a Chinese Aircraft factory from Japanese bomber attacks. When he mentioned the pay would be $350.00 per month, we realized this was a fantastic offer! Most of us were making $72.00 a month. However, the requirements for being hired required technical school, staff sergeant or above, and P-40 experience.

I was only twenty-one years old and didn't really know much about the Chinese or Japanese, but I did have the feeling that the Japanese were the bad guys. I

Curtiss YP37, Langley Field, VA 1940. Assigned to 33rd Pursuit Squadron, 8th Pursuit Group.

S/Sgt. Baisden and Corporal Barouch-Tow target-operators, Martin B-10 Mitchel Field, NY 1941

Honorable Discharge

from

The Army of the United States

TO ALL WHOM IT MAY CONCERN:

This is to Certify, That* Charles N. Baisden

† 6995553; Staff Sergeant, 65th Pursuit Squadron (I), GHQ AF,

THE ARMY OF THE UNITED STATES, as a TESTIMONIAL OF HONEST AND FAITHFUL SERVICE, is hereby HONORABLY DISCHARGED from the military service of the UNITED STATES by reason of ‡ Section X, A.R. 615-360 & WD AGO ltr AG 220.81 (5-23-41) on subject disch.

Said Charles N. Baisden was born in Scranton, in the State of Pennsylvania

When enlisted he was 19 5/12 years of age and by occupation a Machinist Helper

He had Brown eyes, Light Brown hair, Ruddy complexion, and was 5 feet 8 1/4 inches in height.

Given under my hand at Mitchel Field, New York this 24th day of May one thousand nine hundred and forty-one

JOHN E. BARR,
Captain, Air Corps,
Comdg 57th Pur Gp (I), GHQ AF
Commanding.

See AR 345-470.
*Insert name; as, "John J. Doe."
†Insert Army serial number, grade, company, regiment, or arm or service; as "1620302"; "Corporal, Company A, 1st Infantry"; "Sergeant, Quartermaster Corps."
‡If discharged prior to expiration of service, give number, date, and source of order or full description of authority therefor. 16—10506

W. D., A. G. O. Form No. 55
October 10, 1939

Convenience of the Government Honorable Discharge.

AGREEMENT

THIS AGREEMENT made at Gard3n City, N.Y. this 14 day of May, 1941, by and between CENTRAL AIRCRAFT MANUFACTURING COMPANY, FEDERAL INC., U.S.A., 30 Rockefeller Plaza, New York, N. Y., a corporation organized under the Provisions of an Act of Congress known as "China Trade Act 1922", hereinafter called the Employer, and Charles Norman Baisden residing at .... hereinafter called the Employee

WITNESSETH that:

WHEREAS, the Employer among other things operates an aircraft manufacturing, operating and repair business in China, and

WHEREAS, the Employer desires to employ the Employee in connection with its business and said Employee desires to enter into such employment,

NOW THEREFORE, in consideration of the premises and the mutual covenants and agreements herein contained, it is agreed between the parties as follows:

ARTICLE 1. The Employer agrees to employ the Employee to render such services and perform such duties as the Employer may direct and the Employee agrees to enter the service of the Employer who, in consideration of the Employee's faithfully and diligently performing said duties and rendering said services, will pay to the Employee a salary of threehundredfifty dollars United States currency (U. S. $350.00) per month payable monthly on the last business day of each calendar month.

ARTICLE 2. The said employment shall become effective and salary payments to the Employee therefor shall begin as from the date of the Employee's reporting in person to the Employer's representative at the port of departure from the United States (which port shall be designated to the Employee by the Employer), such date to be not later than June 2, 1941, and shall continue for one year after the date on which the Employee shall arrive at the port of entry to China (which port shall be designated by the Employer to the Employee prior to the latter's departure from the United States) unless sooner terminated as hereinafter provided. The Employee undertakes to report to the Employer's representative at the said port of entry immediately upon arrival there. After the expiration of the said term of one year this contract shall continue in effect on the same terms and conditions unless and until terminated by either party on 30 days' written notice.

ARTICLE 3. The Employer agrees to pay all reasonable travel costs to the Employee from Gard3n City, New York to the place in China to which the Employee may be directed to proceed, including:

(a) First class railroad fare and berth, if necessary, by the most direct route to point of departure from the United States plus U. S. $3.00 per day for meals en route.

(b) Actual cost of passport and visas.

(c) Reasonable cost of hotel and meals while awaiting ocean transportation in accordance with travel instructions of the Employer.

(d) Transportation to port of entry to China.

(e) The sum of U. S. $100.00 for contingent expenses en route to port of entry.

(f) Actual expenses from the port of entry to the designated destination in China to be accounted for by the Employee to the Employer.

ARTICLE 4. At the expiration of this contract the Employer agrees to pay to the Employee in addition to the sums elsewhere stipulated in this contract the sum of U. S. $500.00 in lieu of Employee's return transportation costs from China to the United States.

Central Aircraft Manufacturing Company Contract (page 1).

ARTICLE 5. The Employer reserves the right to terminate this contract summarily by written notice to the Employee in the event of misconduct of the employee in any one of the following categories:

(a) Insubordination;

(b) Revealing confidential information;

(c) Habitual use of drugs;

(d) Excessive use of alcoholic liquors;

(e) Illness or other disability incurred not in line of duty and as a result of Employee's own misconduct; or

(f) Malingering.

In the event of termination of this contract for any of the causes above mentioned under (a) to (f) inclusive of this ARTICLE 5, the Employer shall not be obligated to pay the return travel allowance provided for in ARTICLE 4 above nor have any further obligations to the Employee under this contract except that the Employee shall be entitled to the earned portion of his salary computed to the date on which written notice of termination is given by the Employer under the provisions of this ARTICLE 5.

ARTICLE 6. In the event of the total disability or death of the Employee suffered in line of duty the Employer upon receipt of proof of such total disability or death shall immediately pay to the Employee or to the Employee's designated beneficiary as the case may be a sum equal to six months' salary at the monthly rate provided for in ARTICLE 1 hereof. In the event of total disability the Employer shall defray the expenses of transportation of the Employee to the United States. In the event of death the employer shall defray the expenses of a decent burial of the remains. The Employee agrees that he has in good standing a policy of U. S. Government Insurance in the amount of $10,000, Policy No.........................., plus disability benefits and hereby authorizes the Employer to pay the premiums on said insurance during the period of this employment and to deduct the amount of said premiums from the pay of said Employee.

ARTICLE 7. In addition to the payments provided for elsewhere in this contract, the Employer shall provide or cause to be provided to the Employee free of charge:

(a) Suitable living quarters or an allowance of U. S. $30.00 per month in lieu thereof.

(b) Necessary motor transportation.

(c) Suitable medical and dental service.

ARTICLE 8. The Employee shall be entitled to leave with pay as follows:

(a) One month's leave during the first year of employment at such time or times as his services can best be spared.

(b) Similarly in the event of additional service beyond the first year, leave for a period equal to 1/12 of the duration of such additional service.

(c) Sick leave with pay if and as necessary, not to exceed one month in any period of one year.

IN WITNESS WHEREOF the Employer and Employee have executed this agreement in quadruplicate, of which two copies are retained by each party.

CENTRAL AIRCRAFT MANUFACTURING COMPANY
FEDERAL INC., U. S. A.

M. J. H.
Witness

By: Richard Haworth.

Herbert C. Kennedy
Witness

Charles Norman Baisden
Charles Norman Baisden

Central Aircraft Manufacturing Company Contract (page 2).

Aboard President Pierce-14 of the first group to sail for China.

Rangoon, Burma

July 29, 1941.

I hereby volunteer for service in the Chinese Air Force from July 29, 1941 until termination of Current Contract.

During this period I promise to render loyal service to China and to obey the orders of my superior officers and the military laws and regulations of China.

After the termination of my service, I will not reveal any information relating to military matters which I gained during my service.

Charles N. Baisden

B. R. Carney
(Witness)

____
(Witness)

We all signed these when we arrived.

also knew this was a great opportunity to have an adventure and get paid for it, so I decided to sign up. We did have top government approval, but the top brass were very unhappy as they were losing trained personnel.

On May 14, 1941, I received an Honorable Discharge. I turned in all of my field equipment to supply, but I was allowed to keep my uniform after removing the brass buttons from my blouses and overcoat. I also received a U.S. passport which listed me as a metal worker. The State Department would not let me be listed with an armament title.

The same day of my discharge, I went to the corporate offices of the Central Aircraft Manufacturing Company located at Rockefeller Plaza in New York City. There I signed a contract agreeing to be an employee of this company for a period of one year. From New York I traveled to Los Angeles by train where I met up with other mechanics, armorers, and radio men at the Jonathan Club. However, there were no pilots in this first group. From Los Angeles we took a bus to San Francisco and boarded our ship, the President Pierce, for the long voyage to Hong Kong. We had some short stops in Hawaii and Manila.

Once we reached Hong Kong we took a Dutch Packet ship to Singapore where we stayed about ten days at the Raffles Hotel. Because we made ourselves rather unpopular with the locals there, the group following us received a cool welcome. I won't go into the details of some of those antics.

From Singapore we boarded a former German ship for Rangoon. This ship had been turned over to the British after World War I. When we arrived in Rangoon, Burma, we were put up in an old, outdated hotel called Minto Mansions. That night the Chinese gave The American Volunteer Group (AVG) a banquet.

At the banquet I met our boss, Colonel Chennault. I will never forget when I shook hands with him. His eyes seemed to burn right through me. I had the feeling he was trying to determine what kind of people he was getting.

The next morning, our armament chief, Roy Hoffman, called me. "Charley, the Old Man wants you to take the baggage to Toungoo today," he said. "The rest of us will follow by train tomorrow." I set out for Toungoo in an old station wagon with a Burmese driver. Toungoo is located about 150 miles north of Rangoon. We were to operate from a British Air Base, Kyedaw, some eight or nine miles from Toungoo.

I checked in with the British Commander and spent a very uneasy night. I was alone in the barracks and couldn't get to sleep because of all the unfamiliar noises. I kept hearing lizards dropping off the inside of the roof. I kept my .38 Colt under my pillow all night. When morning finally came, I was very relieved.

Later that day my friends arrived by train. We all agreed our living conditions weren't what we were accustomed to. Since we had arrived during the monsoon season, it was rainy, hot, and humid. The barracks were made of teak and bamboo and had thatched roofs. The interiors were open bays with a small room on one end. Our beds were made of teak boards with rope springs and a very thin mattress. It was dark inside with low wattage light bulbs hanging from the rafter. A mosquito net hung over each bed. There was no place for personal items; you had to figure out for yourself what to do with your things. The latrines were one and two-hole "johns" outside without water. The showers were also outside.

Our mess facilities were also bad. They had been contracted to a man named Ussof who regularly fed us bad food. We all agreed he was a professional rip-off artist who was only out to make money off us. He claimed he could not find the kind of food we were used to. On one occasion he served us some revolting water buffalo meat. We told our non-English speaking Indian server that it was called "horse cock." Proud of his newly gained knowledge, he went from table to table

Riffer, Sweeney, McHenry, and Hubler. A long ways from home.

loudly announcing he was serving "horse cock." Later, we were able to get a snack bar going where we could buy toilet articles, cigarettes, and some fine Java beer.

Toungoo had a flat landscape with little shade. Depending on the weather, we were surrounded by dust or mud. Animal life consisted of centipedes, scorpions, short fat snakes, and ill-tempered cobras. When I went to the latrine at night, I made sure I had a flashlight. I usually saw at least two or three scorpions resting close to where I planned to put my family jewels. I also shook out my socks and shoes

Monsoon Season-Burma 1941. AVG Photo archives.

before putting them on. One time someone saw a snake crawl under the barracks. Joe Poshefko threw his GI shoe at it and killed it after four or five of us had emptied our pistols without hitting it.

It was a long haul from the barracks to the flight line, so some of us went into Toungoo and bought bicycles. We could buy a good English bike for thirty dollars. We paid for goods in Burmese rupees. The exchange rate was three rupees to the dollar.

At Toungoo, during the early phase of our operations, there were no aircraft. Instead, they were off-loaded at Rangoon and made ready to fly by a detachment of the AVG stationed there. Once they arrived in Toungoo, the communications and weapons systems were installed and the aircraft bore sighted.

At first, we did not have the proper tools and had to make due with screwdrivers, pliers, and adjustable wrenches that came from the station wagons and trucks at the barracks. And because we were not used to the climate, our work time was limited to afternoons only. Later on, however, we received the correct tools and adjusted to the climate, so we were able to work from dawn until dusk.

Sunday was our day off prior to Pearl Harbor, and one Sunday, William Unger, an armorer from the 2nd Squadron, and I decided to go on a tiger hunt. We hired a Burmese guide who took us to a village near the airfield. Here we spent the night. I

Joe Poshefko and Chuck Baisden—Barracks Kyedaw, Burma, 1941.

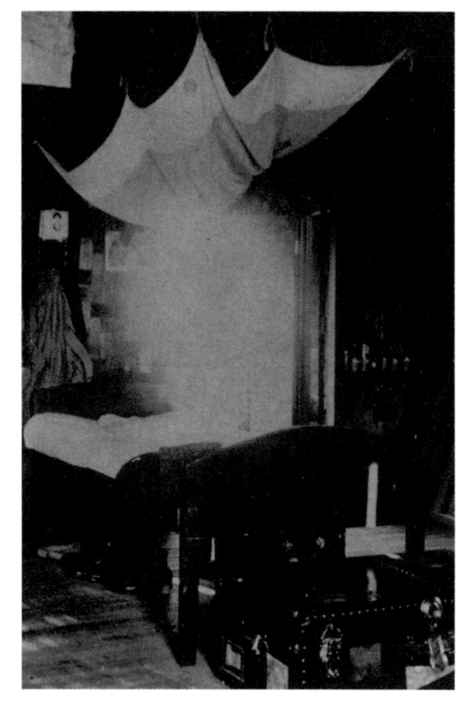

Author's bunk at Kyedaw, 1941.

Taxi accident, Kyedaw, 1941.

Early Tiger Decal, 1942—A Walt Disney creation.

believe we were the first Americans they had ever seen, and they were impressed with my "magic" trick using the hem of a handkerchief and some matches. After putting several unbroken matches in a handkerchief while pretending to have just one, I would have one of them break "the" match. After saying some mumbo-jumbo, I would shake out the handkerchief and an unbroken match would drop out. They were amazed!

The Burmese men and women were small. Almost all seemed to chew "betel nut," a concoction prepared from the nut of the betel palm. After boiling and drying, a small piece of the kernel is cut and placed in a peppery leaf together with a bit of quicklime. This is then rolled into a small ball. When chewed, it turns the gums and lips a brick red and the teeth black. In addition to these colorful features, the Burmese women would cover their faces with a white powder.

I had heard that Winston Churchill's favorite cigar was a Burmese cheroot. I took one apart one time to see how it was made. Instead of tobacco, there was newspaper print on the wrapper. In the village where we stayed before the tiger hunt, there were huge lit cigars hanging from the roof by a string. Anyone who wanted a puff could help themselves.

In the village, we slept on the porch of the headmaster's hut. In the morning they fed us a breakfast of eggs and curry, and then we started out on our hunt, accompanied by at least fifteen of the villagers who acted as beaters. Bill was armed with a 12 gauge shotgun and no.6 shot, some of which he had ringed or cut around the cartridge cases to keep the shot pattern from spreading. I had borrowed a .30-06 Savage bolt-action rifle with hollow point bullets made from our .30 caliber machine gun ammunition.

After our beaters made several drives with enough noise to put any tiger into an afterburner mode, we decided tiger hunting was not our thing. We had succeeded only in disturbing flocks of large yellow-legged birds, and the waist-high jungle growth was full of little white leeches which kept dropping on us as we went by. We burnt them off with our cigarettes. Bill Unger and I paid the village chief one hundred rupees (about thirty dollars). We both agreed we were lucky we didn't actually see any tigers.

My friend, Bill Unger, disappeared sometime after the Japanese started bombing Burma. Some said he joined the British Army, others insisted he joined the OSS, but to this day, I've never found anyone who knew for certain what happened to him.

36

OFFICE OF THE SUB-DIVISIONAL POLICE OFFICER, MAYMYO.

HILL-PERMIT.

Permission is granted to Charlas Baisden to take down his lorry car No. R.E. 4698 from Maymyo to Mandalay to day the 18-11- 1941.

The width of the car is 6 feet 7 inches

Sd. U Ba Choe,
Sub-divisional Police Officer,
Maymyo.

Copy to :—
1. Applicant.
2. D.S.P. Mandalay for information.

Ba Choe
18/11

Sub-divisional Police Officer,
Maymyo.

Hill Permit.

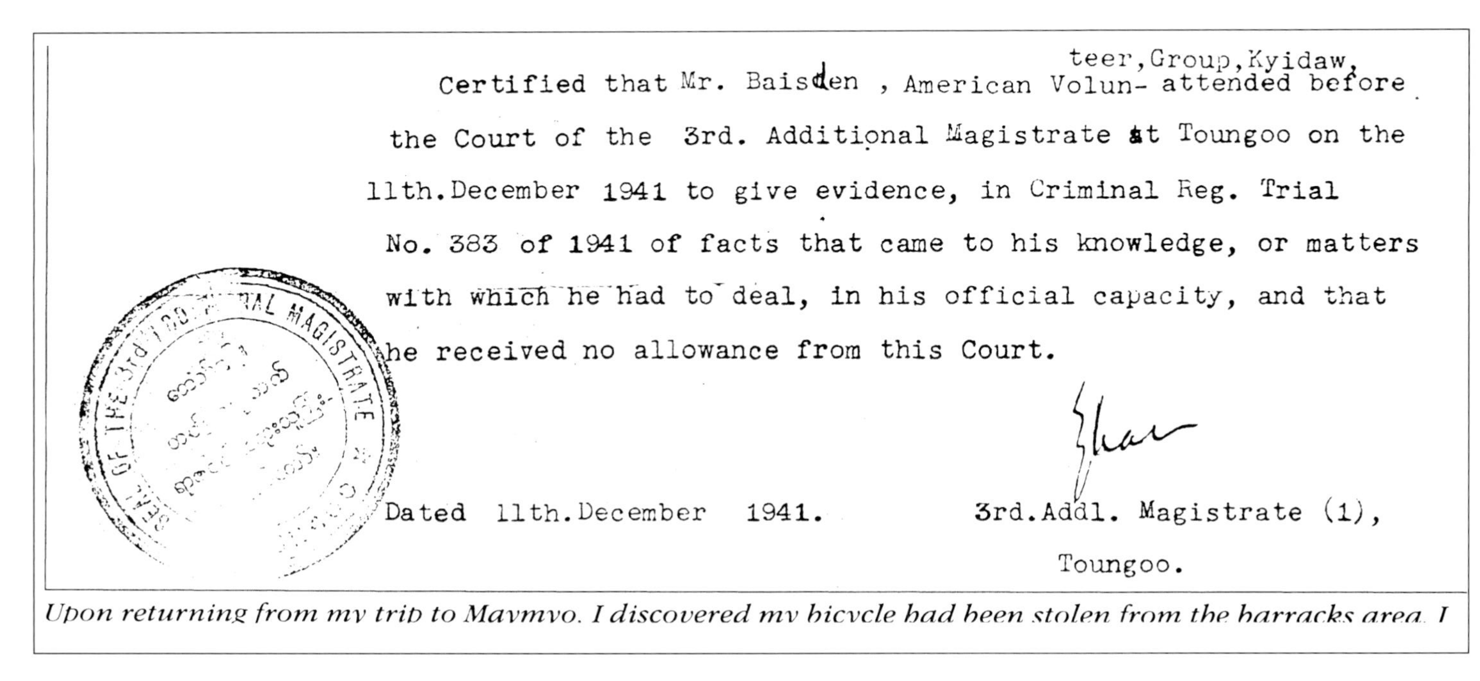

Certified that Mr. Baisden , American Volunteer, Group, Kyidaw, attended before the Court of the 3rd. Additional Magistrate at Toungoo on the 11th. December 1941 to give evidence, in Criminal Reg. Trial No. 383 of 1941 of facts that came to his knowledge, or matters with which he had to deal, in his official capacity, and that he received no allowance from this Court.

Dated 11th. December 1941.

3rd. Addl. Magistrate (1),
Toungoo.

Upon returning from my trip to Maymyo, I discovered my bicycle had been stolen from the barracks area. I

Certificate of trial.

When the AVG first started flying the P-40s there were many problems. A number of the pilots were ex-Navy and had been used to flying slower PBY patrol bombers. They weren't used to landing the P-40. After a number of landing mishaps, Colonel Chennault had a white line painted on the runway. Anyone landing past this line was fined $50.00. There were also a few flying accidents with the pilots being killed, including Max Hammer, John Armstrong, and Pete Atkinson.

An Indian magazine featured a P-40 with a shark-mouthed decal on it. It was from one of the United Kingdom outfits, but someone decided the shark decal would look good on our own aircraft. Stanley Regis, crew chief in the 3rd, painted most of the Hell's Angels squadron insignia. The Walt Disney-designed Flying Tiger decal came out after Pearl Harbor.

On November 16, 1941, Colonel Chennault ordered me to go to Maymyo to pick up 55,000 rounds of 7.9 machine gun ammunition. This caliber was standard for much of the Chinese Nationalist Army, and the lst Squadron of the AVG was being armed with 7.9mm wing guns.

Maymyo is located in the mountains east-northeast of Mandalay, and was a favorite vacation area for the Brits to get away from the heat and humidity of the Burma plains. This trip meant a one-day drive to Mandalay on the Burma Road, then a drive to Maymyo, and, finally, a return to Toungoo. All in all, it was a distance of 600 miles, and part of the way back I had a police escort. I really enjoyed the trip.

However, when I returned to Toungoo, I discovered that my bicycle had been stolen from the barracks area. I immediately reported this to the Toungoo Police, and within two days they had found the bike and arrested a young Burmese boy who had been working in the barracks as a sweeper. I was required to go to the Toungoo jail and identify both the bike and the boy.

The jail was made of concrete, with thick square bars made from teak wood. The floor was bare concrete with a mat for sleeping, a bucket for the toilet, and a single cold water faucet. I really felt sorry for the kid. The police would not return my bike until after the trial, which was on the 11th of December. At the trial, which I attended, the boy got six months of hard labor. After completing his sentence, he was then going to be held for a new trial on other theft charges.

2

Basic Armament

Aircraft Machine Guns

The standard machine gun used by the U.S. Armed Services during World War II was the Browning model of the recoil-operated .30 and .50 caliber M2. These weapons were air cooled and could be fed ammunition from either side using a disintegrating metallic belt. The .30 caliber was able to fire at the rate of 1,350 rounds per minute, weighed twenty-one pounds, and had a range of eighteen hundred yards with a muzzle velocity of twenty-six hundred feet per second. It was also made in different calibers, which included the British .303 and 7.9mm. The 7.9mm was installed in the aircraft of the lst Squadron of the AVG.

The .50 caliber M2 was able to fire at the rate of 800 rounds per minute. It weighed sixty-four pounds and had a range of seventy-two hundred yards with a muzzle velocity of twenty-eight hundred feet per second. In actuality, the ideal range for getting hits in a combat situation was much less, and with the closing ranges where the actual firing took place, the time for rounds fired was in seconds.

The .30 M2 just about disappeared on U.S. pursuit type aircraft during World War II because the .50 M2 took over. The P-40 Tomahawk was the last aircraft I worked on which used four .30 caliber wing guns.

During WWII many manufacturers contracted to make the .50 M2. Most of these were quite satisfactory, but there was one company, the Hi-Standard Company, which gave us a lot of trouble with their weapons. They were crudely put together without much attention to quality.

A new model .50 aircraft machine gun came out around 1947. This was the AN-M3. It looked similar to the M2, but few of their parts were interchangeable. The weapon weighed sixty-four and one-half pounds. Its rate of fire was 1,150-1,250 rounds per minute, and it had a muzzle velocity of 2,840 feet per second with a maximum range of 7,275 yards. The weapon could be fired a maximum of 200 rounds per burst with a barrel life of five thousand rounds.

Gun Sights

The optical gun sight installed in the early model P-40s and used by the AVG were not computing. The sight image projector, or N3A, was mounted below the instrument panel close to the cockpit floor. It projected a sight image of reticle to a rectangular, transparent mirror. Normally, this mirror was attached to the bullet-resistant glass directly in front of the pilot. However, our P-40s were export versions because the RAF had originally ordered them, and they used a different sighting system. Thus, there were no mounting holes pre-drilled in the glass. We had to have a special sight mount made, but there weren't any facilities in Rangoon for drilling the mounting holes. The sight bracket was attached to the pilot's grab bar used for pulling himself out of the cockpit. In addition, a fixed iron ring and bead sight were also installed. Any type of deflection firing had to be estimated by the pilot. This type of gunnery was an art in itself. Not many of the pilots I knew did much deflection shooting. They preferred to bore in behind the target and blast away at close range.

Synchronization

A holdover from WWI was pursuit aircraft armed with machine guns which were mounted on the fuselage and fired between the rotating propeller. This was known as synchronization. The skill of a pursuit or fighter aircraft armorer was measured by how much he knew about this system. Although most people think a machine gun fires automatically once the trigger or firing mechanism is energized, this is not true with weapons designed to fire through the propeller arc. To prevent the bullets from striking the prop blades, a unit called an impulse generator was mounted on the aircraft engine (this was usually installed in the most inaccessible place). As the engine turned, the impulse mechanism generated a forward and backward motion. This motion was transmitted through an impulse wire that was encased in a tube. The impulse wire was attached to a trigger motor, and this motor was mounted to the side of the machine gun receiver side plate. Inward motion of the trigger motor would depress the sear slide, which in turn released the sear and allowed the firing pin to strike the primer of the cartridge. Then the weapon would fire. Only when the sear slide was depressed would the gun fire. The appearance of automatic fire was established as the engine was turning over so many rpms.

In theory, the bullet would be positioned exactly between the propeller blades at the moment it reached the prop arc. Unfortunately, underspeeding or overspeeding the established rpm settings could result in a shot prop. This was also true of a "cook-off" (an overheated weapon) and a "hang fire" (bad cartridge).

When properly maintained, this method of controlled firing worked very well, but there were many drawbacks. By the end of 1942, this system had vanished except for the P-39 and the A-T6. On the newer aircraft, such as the P-40E and P-51, the weapons were installed in the wings, and on the P-38 and jet fighters, such as the F-80 and F-86, they were in the nose.

Harmonization or Bore Sighting

With fighter-type aircraft such as those used in WWII, the theory of bore sighting was much the same. At a given speed, the aircraft will have a nose up or nose down attitude, which is known to armament specialists as "the angle of attack." This was computed in mils plus or minus. The figure was taken from the Aircraft Technical Order and set in the gunner's quadrant when he was bore sighting. I believe the mil setting for the AVG P-40 (H81A) was three mils nose up.

Two methods were used to bore sight. One was to use a one thousand inch range with the target worked out on a ratio and proportion scale. The sight and weapons could be set using this system without actually firing the guns and while working inside a hangar. Many armorers, and I was one of them, did not like this method of bore sighting because it did not give the armorers a chance to test fire the machine guns. In addition, if there was a mistake made in the bore sighting procedure, it was not found until after the pilot flew and fired the weapons. Then we never knew if the sight was off or the guns were off.

The second method used by the AVG was worked out by Colonel Chennault and our armament chief, Roy Hoffman. The aircraft was taken to the firing range where the target was approximately 300 yards from the aircraft (I have never found a record of the exact range, and other AVG armorers vary in their estimate of the range distance). A bar was inserted through the lifting holes in the rear of the fuselage. Then we would manually lift the aircraft so that a truck could be backed under the tail wheel. The aircraft was secured to the truck body by rope tiedowns.

In this second method, an armorer would get into the cockpit and place a leveling bar on the fore and aft leveling lugs, which were on the left side of the cockpit edge. A gunner's quadrant with the correct mil setting was placed on the bar. Then the aircraft tail was raised with a car jack, or else we would let the air out of the tail wheel until the quadrant bubble read level. The leveling bar was then placed across the horizontal leveling lugs, one of which was on the right side of the cockpit edge directly opposite the rear leveling lug on the left side. The quadrant was then set for zero level, and the wings were leveled either with wing jacks or by bleeding air from one of the tires.

With the aircraft battery switch on and the sight projector image showing on the transparent mirror, the center of the reticle was positioned on the target bull's-eye, and the fixed ring and bead sight were installed. The .50 caliber fuselage guns

Bore sighting Chuck Older's aircraft N0.68 at Kunming, China, 1942.

Roy Hoffman-armament chief at bore sight range, Kyedaw, Burma, 1941—* killed in Aircraft crash Chanyi, China, 1942.*

would be bore sighted on the bull's-eye by lifting the belt-feed cover, removing the back plate, the driving spring, bolt stud, and bolt assembly from the weapon's receiver, and then sighting by eye through the bore of the barrel. Adjustments as to elevation and azimuth were made through the rear mountings in which the weapons were seated. The parts were all replaced, and one round was fired at a time by pressing a screwdriver against the trigger motor slide until the bullet hit the bull's-eye.

The wing caliber .30 belt-feed covers were raised, the bolt locked back, and a bore sight tube or indicator was inserted into the chamber of the weapon (this indicator looked like a little periscope). The same adjustments as to azimuth and elevation were made for the .50s. However, the wing guns were sighted as follows: left outboard .30 sighted to a bull's-eye five feet from the right of the center line of the aircraft; left inboard ten feet from the right center line of the aircraft. Conversely, the right side wing guns were sighted the same way to the left of the center line.

Once a rough sighting was made using the eyeball method, the individual guns were fired one round at a time until the bullets hit the target. At that time, the target frames were taken down and a burst of automatic fire was fired from each wing gun by an electrical solenoid mounted on the sides of the .30s and energized through a trigger switch on the joy stick. We found and corrected many malfunctions during this phase of the operation. The .50s were not fired in automatic, as the synchronization system was tied in with the engine running at flying speeds.

This method of bore sighting resulted in a cone of converging fire that would almost disintegrate any aircraft caught in the cone. The pilot had a choice of weapons through a switch mounted on the lower left side of the cockpit. He could choose from .30s, .50s., or all. The wing guns were charged or loaded by a cable system, and the .50s had a manual charger mounted to the sides of the gun receivers.

We found out later that it was much easier for us to charge the wing guns on the ground and lock them back. Charging in six machine guns when at altitude is a chore. All the pilot had to do was give the charging handles a twist to the left and the wing guns were ready to fire. The .50s were still manually charged by the pilot. On most occasions, the pilot assigned to the aircraft would be present for the bore sighting and test firing.

Aircraft Ammunition

In the Army Air Corps, storage, handling, and delivery of bulk ammunition was handled by specialists from Aviation Ordinance Organizations and delivered as needed to the units. In the AVG we kept our ammunition on an armament truck, and the truck went wherever we went. Except for the time I went to Maymyo and picked up a load of 7.9 ammunition, I was not aware who it was who delivered the ammunition to our squadron. Somehow ammunition was always available, although there were a few times when we had to belt up our own. We did this by taking single rounds and inserting them into the metal links with a hand loading belting machine.

Bullets for each cartridge were color-coded on the bullet tips: ball bullets were no color; tracer bullets, which left a red trace in their wake, were red tipped; armor-piercing bullets were black; incendiary bullets, which burst with a white flash on hitting the target, were blue; and armor-piercing, incendiary bullets were aluminum with other color variations. Armorers would also dip bullet tips in lithographic ink in order to identify hits from a specific weapon or to identify who was firing on the same practice target.

In 1939-1940, the standard pursuit loading was four ball and one tracer. In the AVG, we used two ball, two armor-piercing, and one tracer. We tried some .50 caliber incendiaries in our synchronized weapons, but the quality was so poor that we had a number of shot props and stopped using it except in the wing guns of the P-40E aircraft. This was export ammunition made by a well-known manufacturer.

Bombs

The P-40B (H81A-2) had no bomb racks, although there was provision in the wings for carrying two flares. Until we received the P-40E in the spring of 1942 we had nothing to do with bombs. Roy Hoffman and Harvey Wirta did rig one of the B models with some sort of a bomb shackle, but I do not think it was ever used. On the E models, Chinese and French fragmentation (the latter a vintage of WWI) and Russian general purpose demolition bombs were used in lieu of an extra drop tank. The Russian bombs had to be modified as they had only one carrying lug and U.S. bombs had two carrying lugs. U.S. five hundred and one thousand pound general purpose were coming into Kunming around June, 1942.

3

Toungoo to Pearl Harbor

The 3rd Squadron of the AVG armament section was composed of the following personnel: Joe Poshefko, Paul Perry, Clarence Riffer, Keith Christensen, and myself. We were all former staff sergeants from the 8th Pursuit Group, Mitchel Field, L.I., New York, with technical school and P-40 experience, except for Christensen, who was a former Navy petty officer. Another armorer assigned to our Squadron was L.D. Hanley, who was on detached service at Rangoon where the P-40s were assembled. He quit us in March, 1942, and received a dishonorable discharge from Colonel Chennault. We had three Chinese armament officers assigned to the armament section. They worked with all three squadrons.

There were thirty-one P-40s numbered 68 through 99 assigned to the 3rd Squadron. Five armorers were assigned to install and service one hundred and twenty-four .30 machine guns and sixty-two .50 machine guns. However, because of several accidents, we took only eighteen of the thirty-one aircraft into combat at Rangoon soon after Pearl Harbor.

Normally, in an Air Corps Squadron there would be four times that number of armorers. However, in our unique situation we were all very knowledgeable about our work and did not have to have a supervisor checking on us all the time. We all worked together as a team. We also did not have to pull k.p., guard duty, stand inspections, parades, and all the other things which were so much a part of a GI's life.

We had Burmese and Chinese workers under our supervision in the armament shop, and they did most of the actual cleaning of our machine guns. We would install them ourselves and check them over for proper operation. The initial installations of the wing guns required a great deal of filing in the front mounting trunions to get the mounting lugs to set in.

Third Squadron Armorers 1941 and 1981 Reunion.

We spent a lot of time in Rangoon trying to drill through the bullet resistant glass, so another sighting mount was designed by Joe Gasdick, our sheet metal man. When he first made the mounts they were put together by brass rivets. When the mounts were made from sheet aluminum, they came loose due to electrolysis action between these two metals and the high humidity of the area. Finally, they had to be replaced with aluminum rivets. These sights were never really satisfactory.

In October, 1941, we started to bore sight the aircraft. We could only do one aircraft per day because the three squadrons were using just one range. If all went well we could get two aircraft finished. As with any man-made piece of machinery, the end result of how well it functions is directly proportional to the amount of preventive care it receives.

This preventive care is extremely vital to the individual whose survival may depend on how well his equipment functions. A fighter pilot must depend on his armorer because he has little chance of correcting a malfunctioning machine gun once he is engaged in air combat. Any armorer worth his salt knows this. Bomber gunners would get very upset if anyone messed with their weapons. They knew whose butts were on the line.

Air to ground gunnery started in November, 1941. This was when we began to experience a number of gun malfunctions. The synchronized .50s did not give us much trouble, but the wing guns had several bugs. The value of test firing on the range when bore sighting became immediately apparent. It only took a sneeze to cause the optical gun sights to need adjustments. We covered the .50 blast tubes and the wing gun muzzles with masking tape so we could immediately spot a weapon which did not fire when the planes returned. A plane flying over us that was making a whistling sound meant a shot prop and gave an armorer a very queasy stomach.

To me it was a matter of personal pride and satisfaction when a plane I had serviced came back with the masking tape shot off all six guns. My pride increased if the pilot buzzed the field and did a climbing victory roll before he landed. I guess we all had our favorite pilots. Today all AVG pilots are my favorites, however these got a bit more of my attention: R.T. Smith, P.J. Green, Duke Hedman, Parker Dupouy, Ken Jernstedt, Chuck Older, and Bill Reed.

On Monday, December 8, 1941 (in our time zone), the Japanese struck Pearl Harbor and the AVG was put on combat ready status. All of us had the jitters because of the unknown. We were more worried about a Japanese parachute landing than anything else. We took all our airplanes out and lined them up (what an opportunity that presented for a strafing attack!). We gave them a final check to be sure we had not missed any ammunition loadings.

On December 12, the 3rd Squadron went by train to Mingaladon Air Base located outside of Rangoon. Some of the ground crew drove trucks, and our planes came in that afternoon. This was a British base where a squadron of Brewster Buf-

American Volunteer Group-Chinese Air Force, 3rd Squadron-Combat Ready, December 8, 1941, Kyedaw Airfield, Burma.

Mingaladon Airfield—Rangoon, Burma. Flight line after Japanese bombing attacks. Armament bench with open and partially filled bomb craters in foreground. Second Squadron P-40, No.38 in background. Photo taken from our slit trench.

R.A.F. Brewster "Buffalo," Kyedaw, Burma, 1941.

falo fighters, flown by New Zealand pilots, was located. Our quarters were very similar to those in Toungoo—Indian cooks, the smell of curry, and rank latrines. Slit trenches had been dug all around the area, and a British desert-type tent formed our operations area. We performed the usual pre-flight checks and swabbed out the bores of our aircraft, but did little else.

However, on the morning of December 23, 1941, the Japanese sent over a large formation of bombers escorted by several fighters. Our fighters scrambled, as did a large number of Brewsters. Ours took off in one direction, theirs in another. This presented a rather hairy scene. After the scramble, we gathered around our barracks and stood looking up into the sky, just like a bunch of tourists. We had never been under fire before. We heard engines, and high in the sky we saw small silver specks flying in a V formation. Someone started counting, and when he reached twenty-seven, he yelled, "They're not ours. We don't have that many." I jumped into the nearest slit trench about the same time I heard the whoosh, whoosh of the bombs coming down. Suddenly, I became personally acquainted with World War II! The bomb string came from one end of the field to the other, with the explosions becoming louder and louder as they marched their way toward us. I have never been more scared in my life, including the fifty-eight combat missions I flew in B-25s, my gunner duty in B-29s over Korea, and some 2,000 hours as a boom operator in KC-97 tankers. In all those situations, I could do something about what was happening. When you are being bombed, you are helpless. It isn't a pleasant feeling, and I can sympathize with those who had a number of unsettling reactions to these events.

In the air above we could hear the noise of the engines and the bursts of machine gun fire. We saw a parachute coming down with a Jap fighter shooting at him. Sometime later, P.J. Greene came back to the field, battered and bruised. The Jap had not hit him, but I don't think P.J. ever forgave the Japanese for trying to kill him while he was in his parachute. Neil Martin and Henry Gilbert were shot down and killed. None of the ground crew was hit, and when our planes landed, we quickly rearmed them, ready for the next mission.

On Christmas day we were bombed again. Once our planes took off, however, we got ourselves off the field. There were four of us in a Jeep at the end of the runway when a Jap fighter came in on a strafing run. I bailed out into a thorn thicket and got stuck with several large thorns in my leg. One of them broke off and came out of my leg months later, several inches below where it went in. Better a thorn than a bullet!

The RAF had a number of their mechanics killed while working in a hangar during the bombing. P.J. Perry took a bomb fragment in his leg. He took a lot of kidding about this, and many years later at an AVG reunion, he was awarded T*he Order of The Purple Chicken*.

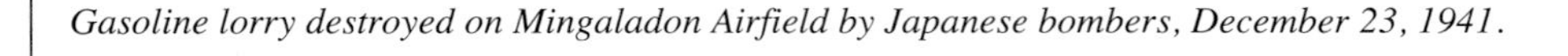

Gasoline lorry destroyed on Mingaladon Airfield by Japanese bombers, December 23, 1941.

3rd Squadron Flight Leader Robert T. Smith (R.T. or Tadpole) just after shooting down 2 Japanese bombers and 1 fighter, Dec. 25, 1941. Mingaladon airfield, Rangoon, Burma, 1941. Bullet holes in R.T.'s P-40 were Christmas Greetings from a Japanese bomber gunner.

Chuck Older and R.T. Smith receive Chinese decorations. Kunming 1942. Courtesy R.T.Smith.

Aces—Ken Jernstedt, Tom Haywood, Chuck Older, and R.T. Smith, the best of the best.

Our pilots really devastated the Japs. R.T. Smith got two bombers and a fighter. His fuselage was shot full of holes. I gave him a cigarette and took his picture. Duke Hedman came in with a one day total of five shot down. He said he shot all his ammo and pushed everything forward and dove. My B-10 tow target pilot, Parker Dupouy, landed with part of his right wing missing after colliding with a Jap fighter.

On that Christmas day our barracks and mess hall were hit and our Indian cooks disappeared. Someone gave me a cold liver sandwich and a bottle of warm Australian beer. This was Christmas dinner!

On December 29, we left Mingaladon for Kunming, China.

4

Kunming to Magwe to Loiwing

In Kunming we stayed in the dormitory of what had been a university. Keith Christensen and I shared a room on the second floor. We had a houseboy to keep the place clean and get our laundry done. He also made himself useful in other ways. He got us a clay charcoal pot to provide some heat for the room as it was cold in Kunming. We almost asphyxiated ourselves one night when we forgot to open a window and the charcoal heater used up most of the oxygen in the room.

The Chinese really tried to do everything in their power to make us welcome. The food was the best we had eaten in a long time. There was a bar in the hotel, but they had the worst beer ever made. I think it had soured. It was a twenty minute drive through the city of Kunming to the airfield. The city itself was in bad shape, and the people were dressed in rags. There was a place called "thieves market" where you could buy back the windshield wipers and hub caps that would disappear from your vehicle if you were not careful. If you couldn't make it to the airfield when an air raid alert sounded, there were hundreds of grave mounds surrounding the city which made an ideal shelter.

We were busy rebore-sighting our aircraft, but, except for occasional Japanese bombing alerts which were always false, we had a lot of spare time for sightseeing. We went to the Nan-Ping restaurant for egg-fried rice and we bought lots of souvenirs. Once we went to an opium den where we received quite a shock at seeing these unfortunate people lying on filthy bunks sucking on their pipes and staring blankly. The guy who ran the place wanted us to pay him for letting us see it, but I opened my coat and showed him my .38 special nestled inside my shoulder holster, and he decided to waive any charges.

There was a rumor we were going back to Burma. Around the middle of March 1942, I was called into the office of Squadron Leader Avrid Olson who asked me

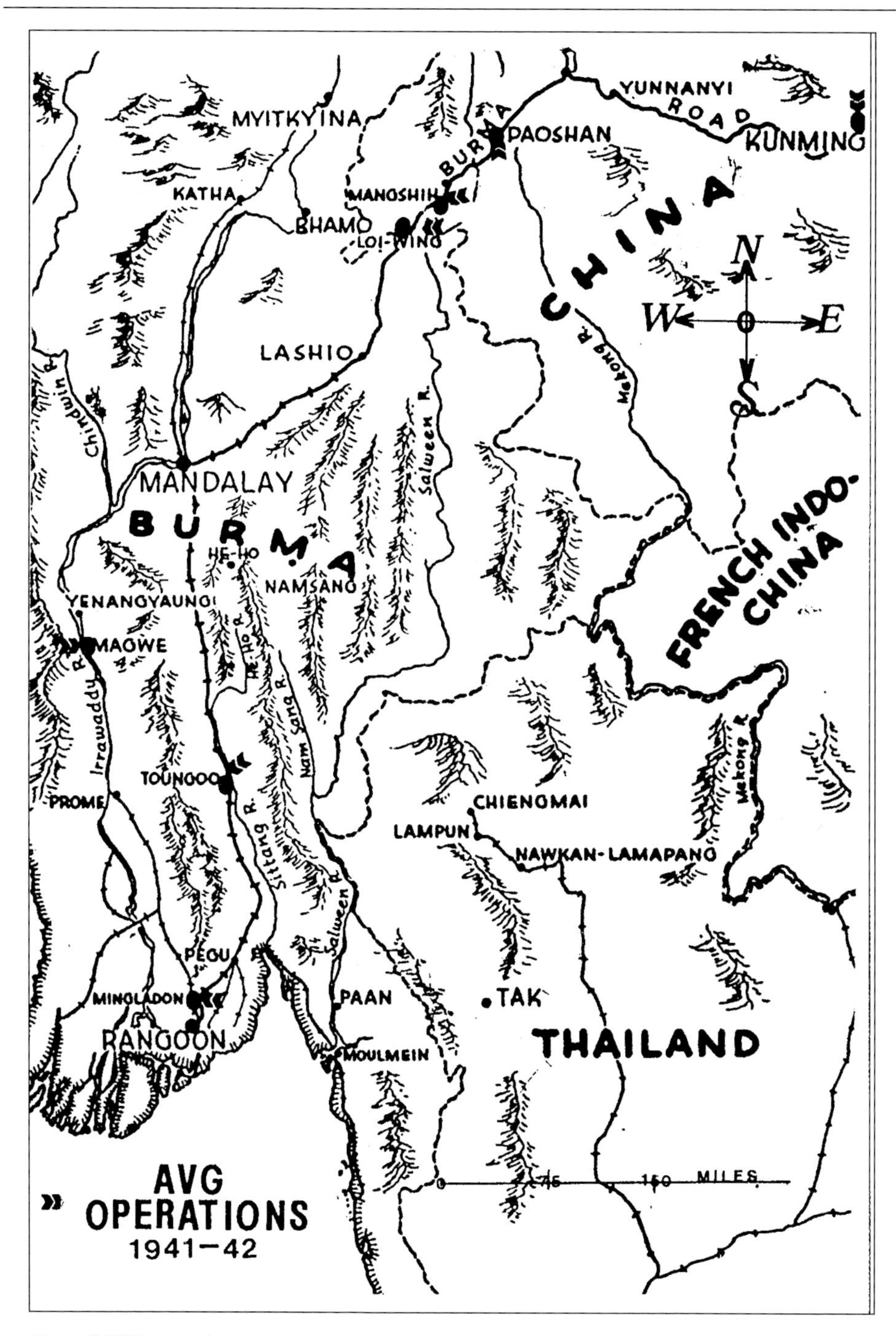

Map of AVG operation area.

my opinion about acquiring some ground defense weapons to take with us to a place called Magwe. He showed me a number of boxed Bren guns of .303 caliber and a couple of old Browning .30 water-cooled machine guns. These were really antiques. He said they had come off the U.S. *Panay*, a gunboat the Japanese had bombed and destroyed at Chungking in 1937. We settled for four Brens and two water-cooled .30s. Olson didn't want the latter, but I promised I'd take care of them.

We took off for our new base on a C-47 which almost crashed on take-off. We had to return to Kunming and leave the next day on a China National airline C-47. We finally made it to Magwe, but that first take-off was as close to losing the 3rd Squadron support personnel as I ever wanted to get.

Author manning German 7.9mm Maxim machine gun on anti-aircraft mount. Kunming 1942.

Magwe is near the Yenangyaung oil fields and is southwest of Mandalay. It was very desert-like, with lots of scrub trees and some rolling flat ground. It was a British Air Base with an asphalt runway. We were quartered in the town some distance from the field, and we had the usual British-type desert tents for our operations and alert facilities.

Since our aircraft came to the base fully loaded, I had plenty of time to set up the water-cooled .30s. I placed them in an L-shaped slit trench near the ops tent. I had to belt about 300 rounds of our aircraft ammunition using an old-fashioned belt-loading machine, much like a miniature sewing machine. I added lots of tracers to the ammunition. Three of the Bren guns were assigned to the ops tent, and I kept the remaining one for my own use. After working on the .30 water-cooled guns, I had some misgivings if they would work without a lot of stoppages. The

Clearing a stoppage on a 7.9mm Maxim machine gun, Kunming 1942.

web cartridge belts were very old, and I doubted they would feed properly. I was not able to test fire them.

The Bren guns were the automatic rifle used by the British army, and in my estimation they were better weapons than our own Browning automatic rifles. The weapon was a Czechoslovakian invention that took a 30 round clip of .303 ammunition. It came complete with a mount that could be used for fixed defense against ground targets and could be easily converted to fire on low flying aircraft. If it was removed from its mount, it could be sling-carried and fired on the move. It was also equipped with a bipod. A new barrel could be installed in seconds, and the drum rear sight made range setting very easy. I carried the Bren around for some time and had many occasions to fire it. I found it very accurate.

One day Ken Jernstedt and Bill Reed flew a mission into the Moulmein area. I had obtained a cluster of British incendiary bombs and figured out a method of placing three incendiaries in the flare rack. I talked Ken into trying them out. I also installed some Canadian concussion hand grenades in Bill Reed's flare racks. Ken got a Jap aircraft with the incendiaries, but I never learned if Bill hit anything with the grenades.

The town of Magwe offered nothing in the way of entertainment, and the closest place to get a beer was at the Yenangyaung oil field club some twenty or thirty miles from Magwe. Several of us went there one night and found there was beer, but there was also a scotch liquor I had never tried called "Drambuie." It was cheaper than beer, and I made the very bad mistake of drinking too much. The next day I had the worst hangover of my life, I think. I felt so bad I decided to try walking to the airfield. On the way I heard the noise of aircraft engines and realized they were Jap bombers. Suffering from a headache while being bombed and taking cover in a ditch that someone had already used for a latrine made for a really unhappy day for me.

The next day when we had a bomb alert, I got off the runway and set up my Bren in a shallow ground indentation. There were some scrub bushes for cover. I was prepared if a fighter came in strafing the field. The bombers did bomb the field, but flew past me, so I packed up and went back to the ops tent. About the time I got there another wave of bombers were coming in from another direction. I grabbed my Bren and jumped into a Jeep driven by "Buck" Rogers. We drove to the same place I had been before. The bombers dropped their load, but I saw a Jap fighter coming in to strafe. He was within one hundred yards of me, but all I could do was hold the Bren up and fire a full clip at him because I had left the anti-aircraft mount on the other Jeep with all my extra ammunition. He fired his machine guns, dropped some anti-personnel bombs, and flew off.

On this raid we suffered several casualties. John Fauth, a crew chief and good friend, was badly wounded and died the next day. He was buried at Magwe. Also a pilot, Frank Swartz, 2nd Squadron pilot from Dunmore, Pennsylvania, a borough

The end for R.T. Smith's No. 77 Magwe, Burma, 1942.

next to my home town of Scranton, was wounded. He died in a hospital in India about a month later. I visited his mother when I returned to the States in September, 1942. Wilfred Seiple was also hurt. He received a concussion.

Many of our planes on the ground were damaged or destroyed. We soon received orders to evacuate the base. Keith Christensen and I removed the machine guns from several aircraft, including R.T. Smith's No.77. It had crash landed on the runway due to engine failure on take-off. Dick Rossi had been the pilot, as R.T. was on a ferry mission to Africa to pick up our new P40Es.

AVG P-40 destroyed by Japanese air attack at Magwe, Burma, 1942.

Ground crew barracks at Loiwing, China, 1942.

AVG operations shack at Loiwing Air Field 1942.

Once we had loaded up all our armament gear, we took off on the Burma Road for Loiwing, which is just over the Burma border into China. We spent one night in Lashio, where I got into an argument with some general just because he was not pleased when he saw us trying to take the innerspring mattresses out of the hostel there. Some of the matresses did make it to Loiwing.

Our quarters at the new base were at the end of the runway on top of a hill. The tin roofs were painted in camouflaged colors. Bamboo and mango trees grew all around the area. There was a factory some distance away that repaired and built aircraft for the Chinese. The operations shack was made from bamboo and thatch. We put a sign in front which said, "Olson and Company, Exterminators, 24-hour Service." The control tower was nothing more than a platform erected on stilts some fifteen feet above the ground.

Bill Towery took over the cooking arrangements for us. He scouted all over the country, buying whatever food he could find. Once he was sold duck eggs, but he thought they were chicken eggs. He made good pancakes, cooking them on a grill which we made from the back armor plate of one of our destroyed P-40Es. He failed to remove the green primer paint from the "grill," and we all had a case of "Towery's revenge."

By this time we had our aircraft armament working extremely well. There wasn't much to do other than the usual pre-flight and swabbing out the gun bores. A couple of misplaced RAF armorers had joined us when we left Magwe, and they helped out a lot. They had been left behind when their outfit pulled out.

I remember some events from this period. The Chinese workers at the aircraft factory challenged us to a basketball game, and they cleaned our clocks. Also, "Fearless" Freddy Hodges got married. We arrived at the factory club house a bit late for the ceremony. On the way back to our own quarters we were stopped by a Chinese Army roadblock. To keep our station wagon from going anywhere they used a very effective method. One of the soldiers put his hand inside our car while holding a grenade that looked like an old wooden potato masher, with the arming ring looped around his finger. They were looking for a Jap crew member who had bailed out of an observation plane. They were very serious about the whole affair, and we cooperated one hundred percent.

Two new P-40E models arrived at the base. They were a big improvement over our older model, with six .50 caliber wing guns, a belly tank, a place under the wings for small fragmentation bombs, and an improved gunsight.

On April 8 we had an alert. Some of our guys took off to intercept an observation plane, but it escaped. Several years later on a trip to Japan, I met Masakazu Shimizu, a retired colonel from the Japanese Self Defense Force, who at that time was in charge of the weather units in Burma. He told me he was the one flying the weather recon missions to Loiwing.

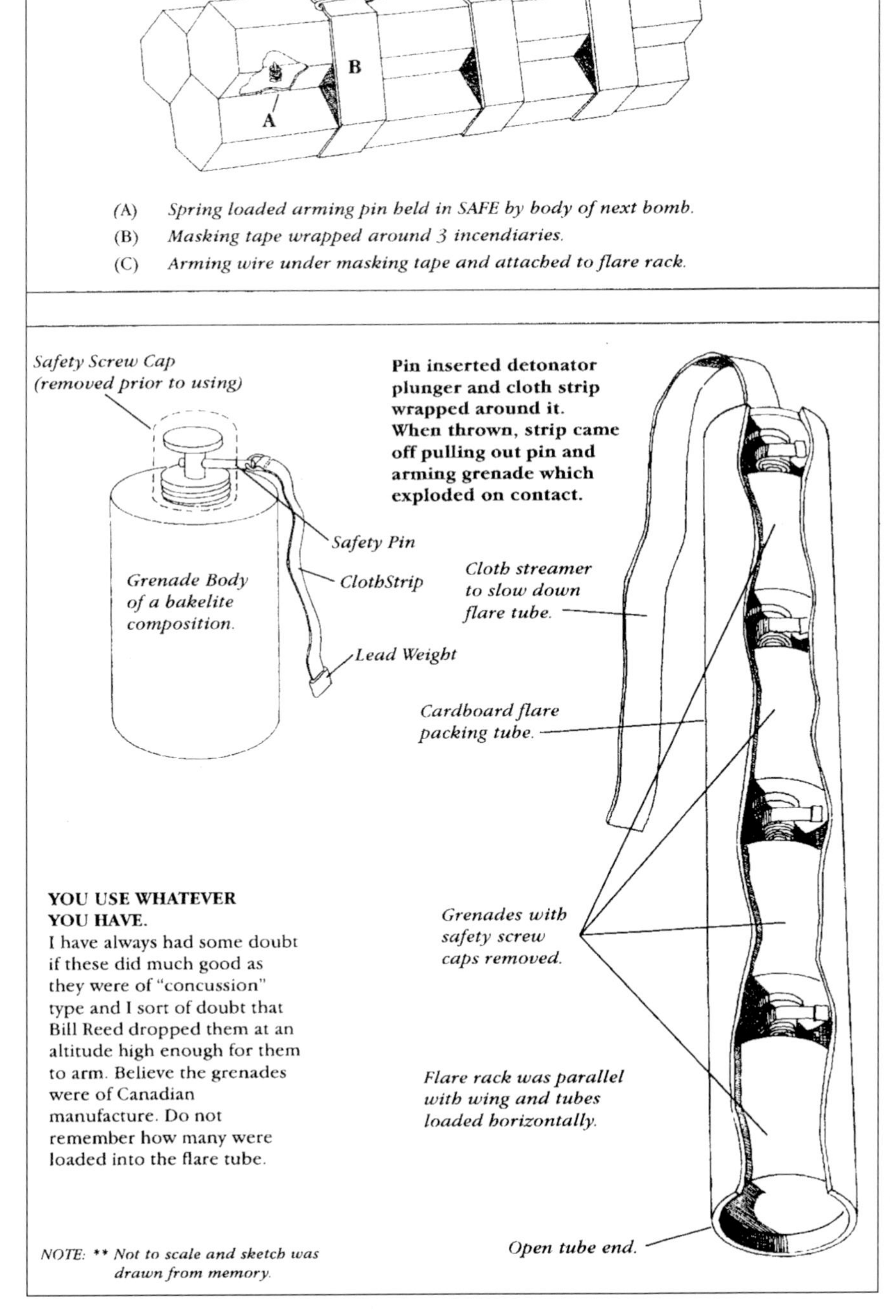

Drawings of incendiary and grenade bombs developed by author to use in flare racks of Ken Jernstedt and Bill Reed's P-40s.

Replacement Curtiss P-40E aircraft at Kunming 1942.

A replacement P-40E destroyed by Japanese strafing attack on Loiwing in April 1942. The aircraft landed in the afternoon and was destroyed the following morning. Never saw the 1st day of combat.

A few days after the alert, Jap fighters hit us with no warning. I was just getting out of bed when we heard their engines and machine guns. I tried to get out the door, but our radioman, Harvey Cross, who was a pretty big guy, knocked me to the floor. Several of our planes were hit, but the damage was surprisingly small. One of our crew chiefs who had been running up his engine thought one of our guys was firing his own guns until his instrument panel disintegrated. It was a miracle no one was injured.

During that same day another air alert was sounded. After our aircraft took off, Joe Poshefko and I cleared the field and sat on a hill overlooking the runway. Suddenly, Jap fighters began strafing the airfield. We could see smoke where once our P-40Es had been parked. Then our own fighters hit the Jap fighters, and we were witness to a low altitude dog fight. The sky started to rain Jap planes. I had my camera and took one picture of a Jap fighter that was diving straight into the ground. He was not on fire, but evidently had been killed in the cockpit. Later on I found it was a victory for Avrid Olson. I have no idea how many planes were shot down, but I'll never forget that afternoon and how proud I was of our guys.

The picture I took of the Jap crashing is, I believe, the only actual photo of a Japanese plane shot down by the AVG. We had no gun cameras in our aircraft. Combat film seen today was made by the China Air Task Force or the 14th Air Force.

Joe Poshefko and I went down to where the Jap had crashed. It was about two hundred yards away from us. The ground was very hard and the plane was completely destroyed and burning. The pilot had been thrown clear, but his body was burnt beyond recognition. There was just a bit of his flying suit still attached to his trunk, and there was no head, arms, or legs. I picked up an instrument gauge, a .50 caliber machine gun cartridge, and the receiver part of a machine gun with the synchronizing trigger motor still attached. The .50 cartridge was new to me. It was an explosive bullet.

Several of us armorers went to the operations shack and found Colonel Chennault there talking with the pilots. I went to the armament room in the same building and began working on a Lewis machine gun. This weapon took a drum of fifty .303 ammunition. While I was fooling with it, I accidentally let a round go through the roof. Avrid Olson promptly chewed me out. A few minutes later, we saw what we thought was a Jap fighter coming in low on a strafing run. I set the Lewis gun on one of the cross members of the control tower, but then decided that discretion was the better part of valor and dove for cover. In fact, everyone found cover, including Chennault. We soon learned that Eric Shilling had decided to give us a buzz job in an old Army trainer. It looked like a Jap fighter from a head-on approach. The smile on Eric's face as he taxied in soon changed when Chennault started chewing him up one side and down the other (I've always regretted not putting a few holes in that plane).

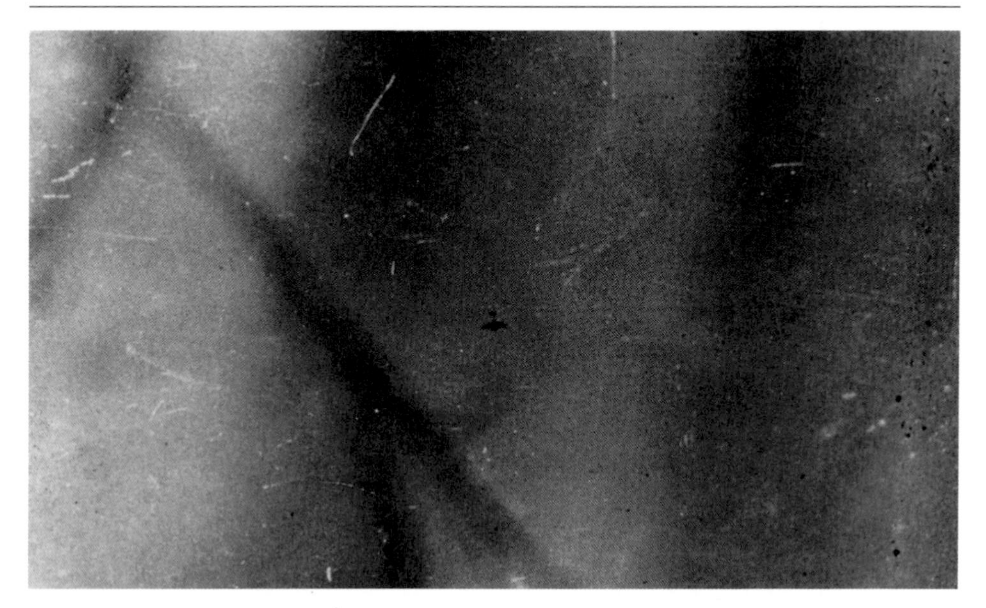

A Japanese Army Nakajima KI-43 Hayabusa, *or Pergrine Falcon, seconds before crashing after being shot down by Sqdn. Ldr. Avrid Olson. Loiwing, April 1942. Called an Oscar by Allies.*

Part of a Japanese 12.7-mm. Type 1 machine gun from wreckage of Oscar in previous photo. The weapon was synchronized to fire through the propeller as noted by the trigger motor on the receiver side plate. There were 2 Type 1 machine guns on this aircraft using impact exploding type ammunition.

A number of aircraft arrived from Kunming, and we began having trouble with some of the wing guns. One squadron had 7.9 caliber machine guns and the ammunition was not interchangeable, even though it was similar in appearance. With several scrambles going on every day, some of the planes did get the wrong ammo, but we only caught it when we tried to charge the guns in on the ground.

Clarence Riffer set our armament truck on fire when he struck a match to check the gas gauge. The gauge was located under the front seat and the gas cap was off. Someone jumped in the ammunition truck and got it clear while the rest of us put out the fire.

There were many refugees. Two big transport planes came and took out as many as they could carry. Since they could only take hand-carried baggage, we received all kinds of goodies they had to leave behind. Joe Poshefko was presented a little English sedan with a sunroof. We had fun with that little car. Some of the guys received pistols and suitcases full of clothes.

During a maximum effort mission, we loaded some of the P-40Es with six small fragmentation bombs. They were manufactured by the Chinese and weighed about thirty pounds each. The detonators, which went into the nose, were blank cartridges about .38 caliber. I've heard this mission stopped the Japanese advance up the Burma Road.

Author—Kunming, China, 1942.

Author's Thompson sub machine gun.

5

The Burma Road from Loiwing, Mengshih, and Kunming

Since the rainy season was beginning and the Japanese ground forces were getting close, we loaded up our gear and trucked up to a place called Mengshih, about four hours by truck. It was just a dirt landing strip that had turned to mud with the rain. It was a terrible place to service our aircraft.

Bob Brook landed, and when I started to rearm his airplane, I found he had fired very little ammunition. He asked me if I could figure out how many rounds he had fired. The best I could figure was not over thirty rounds from each wing gun. He said he had been in some bad weather and was just coming out of a cloud formation when he found himself tail end with three Jap fighters. He got right on the end of the Jap's tail and gave him one burst. It knocked him out of the sky. Then Bob dove out. A puzzling thing to me was that I found a .30 caliber bullet lodged in one of his ammunition containers. It was one of our own.

We did not stay in Mengshih but a few days. Instead, we headed for Kunming. A CNAC plane came to pick up ground people, but I decided to finish up the Burma Road trip by truck. Clarence Riffer drove, and I rode shotgun. The first night we found out we had no food except a box of crackers and crab claw meat we found under the front seat.

The next morning, while Riff was trying to start a fire with some damp wood, he got a tin can of gasoline and proceeded to pour it on our smoldering fire, which set the can on fire. He quickly threw it away, but it landed on my leg, and I still carry the scars of that incident. Further up the road our truck quit on us. It had a broken gas line going into the fuel pump. However—and I know this is hard to believe—we found a carton of airplane model cement under the front seat, so we wrapped a roll of bandages around the broken fitting, applied the glue, and it worked fine.

We ran into a line of traffic and found that a Chinese truck had bogged down in a ditch. A huge crowd of Chinese had gathered around the blocked traffic, but they were doing nothing. Riff and I found a pile of telephone wire in a nearby ditch, wrapped a mess of it around the mired truck axle, attached it to another truck, and pulled him out. As we were pulling him out, another American came up to us and thanked us. He turned out to be Dr. Seagraves (the Burma surgeon), a real fine person.

We had trouble whenever we attempted to pass trucks driven by the Chinese. Along the road, there were a number of deep gorges. On one occasion a Chinese driver tried to dump us into one of these gorges of some one thousand feet when we tried to pass him. However, when it happened again, I put out a burst from my tommy gun across his bow. We didn't have any more trouble after that. I found out later they would lose face if you passed them. I think both Riff and I were more concerned about losing our own butts than about them losing face.

We were driving after dark one night and the road was very narrow. Suddenly, a large tiger jumped across the road in front of us. It seemed like his tail was still on one side while his head was on the other side as he stretched across the road before us. We both agreed it would be best for us to drive through the night instead of spending the night on the road. We arrived in Yunnanyi the next day.

Riff and I decided to stay an extra day there since we had been told it was great deer and pheasant country. We only hunted part of the day and never saw a deer. It wouldn't have been fair anyway, because Riff had the Tommy gun and I had my Bren. My burned leg was bothering me, but I had a first-aid kit I had picked up while at Magwe from a downed British Hurricane fighter. I bandaged my leg using some burn ointment called Gention Violet. It turned everything I wore and my skin a very deep purple, and it took forever to wear off.

When we finally arrived at Kunming, we took our truck to the motor pool and told Mike Wakefield what had happened. He laughed at our repair job, but had a very hard time trying to cut off the glue-soaked bandages. Then I went to see Dr. Rich about my leg. He fixed me up with more Gention Violet and, except for having my tooth filled by Dr. Bruce at Toungoo, this was the only time I ever had to go see our medics for any kind of treatment while in the AVG. I consider myself very lucky, because many of the fellows had dysentery, malaria, jaundice, and other ailments received from places they should have stayed out of.

One day while I was working at our armament building cleaning some of the weapons we had originally taken to Magwe, Chennault drove up in his staff car. He explained that some Chinese Nationalist soldiers had deserted and were holed up in a village across the rice field from where we were working. The Chinese military police had gone in to pick them up, but a fire fight had broken out. We had all heard the rifle fire, but simply thought it was range practice. Chennault wanted me to take a Bren and guard our ammunition bunker. He even told me to shoot anyone trying

Guarding ammunition dump at Kunming with a Bren gun.

Chinese flown-Russian built Polikarpov *I-153. 1000hp-9cylinder radial. Top speed about 250mph. Armed with 4 synchronized 7.62-mm* ShKAS *machine guns. These also fired an explosive type bullet.*

The men who sat in the cockpits were the real Flying Tigers. 3rd Squadron pilots HELL'S ANGELS.
Kneeling, from left to right

Reed, W.N.	*10 confirmed victories - killed in China 1944*
Laughlin, C.H.	*5 confirmed victories*
Overend, E.F.	*5 confirmed victories*
Haywood, T.	*5 confirmed victories*
Older, C.	*10 confirmed victories*
Bishop, L.S.	*5 confirmed victories*
Foshee, B.C.	*0 killed in bombing raid Paoshan 5/4/42*
Hedman, R.P.	*5 confirmed victories*
Donavan, J.T.	*0 K.I.A. Hanoi 5/12/42*
Cavannah, H.J.	*0*

Standing, left to right

Greene P.J.	*2 confirmed victories*
Dupouy, P.S.	*3 confirmed victories*
Groh, C.	*2 confirmed victories*
Adkins, F.W.	*1 confirmed victories*
Raines, R.J.	*3 confirmed victories*
McMillan, G.B.	*4 confirmed victories*
Olson, A.E.	*1 confirmed victory*
Smith, R.T.	*8 confirmed victories*
Jernstedt, K.	*10 confirmed victories*
Brouk, R.R.	*3 confirmed victories*
Schilling, E.	*0 confirmed victories*
Hodges, F.S.	*1 confirmed victory*

The guys with the dirty fingernails and greasy khakis made it all possible. 3rd Squadron ground crew HELL'S ANGELS.

Bartling, W. E.
Bond, C.R., Jr.
Burgard, G.T.
Hedman, R.P.
Hill, D. L.
Jernstedt, K. A.
Lawlor, F.W.
Little, R. L.
McGarry, W. D.
Neale, R. H.
Newkirk, J. V
Older, C. H.
Overend, E.D.
Prescott, R.W.
Rector, E. F.
Rosbert, C. J
Rossi, J. R.
Sandell, R. J.
Smith, R. H.
Smith, R.T.

AVG Aces.

HEADQUARTERS AMERICAN VOLUNTEER GROUP
Office of the Commanding General

Chungking, Szechwan, China,
July 4, 1942

TO WHOM IT MAY CONCERN:

This is to certify that the bearer Charles N. Baisden Armorer, having honestly, faithfully, and diligently fulfilled the terms of his contract with the Central Aircraft Manufacturing Company and served with honor and distinction in the American Volunteer Group from July 28 1941, to June 27 1942, is HONORABLY DISCHARGED from the American Volunteer Group.

It is requested that all persons concerned give all possible aid to the bearer in his journey to the United States.

C. L. CHENNAULT,
Brig. Gen. U. S. A.,
Commanding A. V. G.

Copy of Chennault's request for assistance. It was of no help!!

to break in. I spent a real uneasy afternoon perched on top of the bunker. Thankfully, I didn't have to shoot anyone.

We had a gunnery and bombing range on a hill some distance from the airfield. The range itself was built from stones laid up to form a wall about three feet high. The first time we used the range for bombing practice, Tom Jones flew a practice dive bombing mission. We loaded him with six Chinese practice bombs that were made from ceramic material and filled with white powder, probably rice flour. When the bombs hit they left a white mark. However, we couldn't keep the Chinese locals out of the target area. They would run into the target ring looking for spent cartridge cases because they got a reward for turning in the empty brass cases. This time, however, there were no brass cases, but they were on the range looking for them anyway. Finally, I had to shoot my pistol over their heads before they got the message. Tom made several runs. Then, suddenly, he flew his aircraft right into the

American Volunteer Group
Chinese Air Force
This is to certify that
has been a Member of the
American Volunteer Group
from 1941 to 1942 and is hereby
Honorably Discharged
Character
Group Commander

Honorable Discharge AVG.

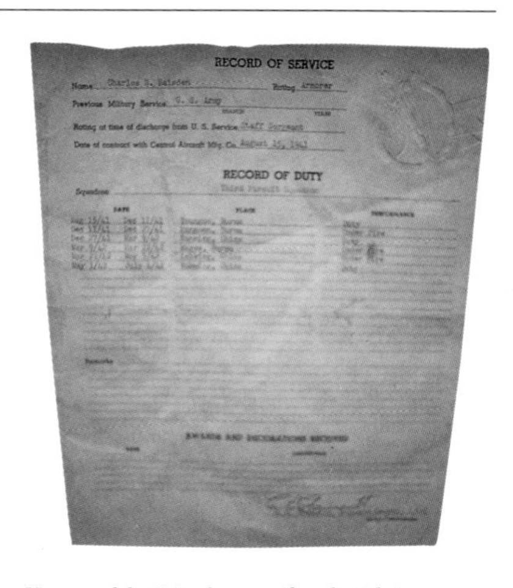

RECORD OF SERVICE

RECORD OF DUTY

Honorable Discharge (back side).

China Service Medal.

Chinese Yaun-front and back. It took many yaun notes to buy much of anything.

target. I think he had target fixation, but we never knew because he was killed instantly. It was a very tragic and unhappy day for all of us.

We were also working with the Chinese Air Force at this time. They started checking out in our P-40s. The pilots chosen for checking out had a lot of flying time, and except for a ground loop or two, most of them did quite well. Some of them checked out in the Republic P-43A1 fighters, and some of them crashed after take-off.

I remember a Russian-built Polikarapov 1-153 that took off with a jury-rigged bomb slung under the fuselage. The plane had only taxied a few feet when the bomb exploded. On another day one of the Chinese airmen was flying over our area when suddenly the plane's tail started to pitch up and down like a yo-yo. The pilot, who did not have his safety belt fastened, was tossed out. However, he popped his chute and landed with only a broken leg. It was a comedy to everyone but the pilot.

Chennault was supposed to receive some B-25 bombers, but the top brass in India decided they would raid Lashio on their way to delivering them in Kunming. However, they got jumped by Jap fighters, then got lost, and finally, ran out of gas. When two arrived one had a dead radio operator. Someone said there were actually supposed to be five bombers delivered. I had never seen Chennault so mad. He knew nothing about the Lashio mission or the Doolittle raid on Tokyo. These people could have received a lot of help if he had been able to put the AVG and Chinese radio network into the picture.

We began to hear a lot of rumors about the Army Air Corps taking over the American Volunteer Group, and some of the Air Corps people started to show up around Kunming. We finally had a meeting with General Clayton Bissell, who told us we should be inducted into the Air Corps in China. There is evidently a law that prevents an American citizen from enlisting or being drafted in the U.S. military service when in a foreign country, but a citizen can choose voluntary induction. Bissell said if we refused voluntary induction, we would not get any help getting back to the States. Also, he told us the draft board would meet us at the dock when we arrived back there.

His attitude of "take it or leave it" was one of the main reasons the majority of the AVG left China. If General Chennault had asked me I would have stayed (Chennault had been promoted to Brigadier General in April, 1942). Thirty-eight of the AVG went into the U.S. military. Two hundred-eighty left with thirty staying an extra ten days before we disbanded July 4, 1942.

6

AVG Days End - Going Home

I left with the first group in June, 1942. We had been the first to arrive in Burma, and we all had a 30-day leave coming. The Air Corps brass made leaving difficult, trying to prevent us from getting transportation, but the Air Transport Command crews were nice to us and arranged transport out of China by telling about twenty of us to be at the end of the runway at a specific time. Then we just got on board the plane and left.

Our C-47 only got us as far as Karachi, India, however, and here is where we hit a brick wall. General Chennault had given us a request in writing for help in returning home, but it made no impression with officials. Some of the officers in Karachi would have liked to help us, but the word was out, and they could do nothing without getting themselves in trouble.

Some of the fellows decided to go to Iran or Iraq and work on the trucks going to Russia on the Lend-Lease program. Some of the pilots in our bunch were hired by the China National Airways Corporation (CNAC). Others, including myself, went to see the U.S. Consul. Here we were told to go to Bombay and wait further notice. Carl Buglar, J.E. Terry, and I went by train to Bombay. There we rented an upstairs apartment which included our meals. For the first time in ten months we had a chance to take a little R & R. I bought a gallon of Coca-Cola syrup to make rum cokes for $18.00, since soda water cost next to nothing. Every Wednesday was steak and kidney pie day, but I soon hated the smells coming from the kitchen.

Around August 1, we boarded the *Mariposa*, a former Madison Line cruise ship converted to a troop transport. It was returning to the States almost in ballast as the only passengers were a few civilians, some missionaries, and former members of the AVG. Some of the AVG had boarded the ship in Karachi. The day we sailed

IMMEDIATE ACTION

WAR DEPARTMENT
HEADQUARTERS ARMY AIR FORCES
WASHINGTON

AAF 342.06 (AFPMP-2)

August 11, 1942.

SUBJECT: Reenlistment of former enlisted men who have served with the American Volunteer Group.

TO: Commanding Generals,
all Air Forces
all Army Air Forces Commands
Commanding Officers,
all stations of the Army Air Forces in continental United States.

1. A number of former Air Forces enlisted men who served with the American Volunteer Group in China have returned to the United States and it is especially desired to have these men reenlisted in the Army Air Forces.

2. A letter has been written to each individual suggesting that he contact the nearest Air Force station commander for full information regarding reenlistment in the Army Air Forces. It is desired that Air Force station commanders interview such of these men as may report to their station and furnish this headquarters, direct, by radio (Attention: AFPMP, Enlisted Section), the following information:

a. Full name of man interviewed
b. Home address
c. Recruiting officer it is desired be authorized to enlist the man
d. Grade it is recommended the individual be enlisted in

3. A number of these men were discharged as privates, but in view of their experience in the AVG, it is believed that they should be enlisted in a higher grade up to and including Staff Sergeant, depending on their experience which will be determined by interview. Enlistment in a higher grade is desired in order to compensate for the grade that they would probably have attained had they remained in the Army Air Forces. In no instance will these men be recommended for reenlistment in a grade lower than that held at the time of discharge from the Army Air Forces. Upon enlistment these men will be assigned to a Fighter Unit.

By command of Lieutenant General ARNOLD:

Wm W. Dick
WILLIAM W. DICK
Colonel, A.G.D.
Air Adjutant General.

DISTRIBUTION: "A"

3 - 1309

Copy of Immediate Action Order from Lt/General Arnold. Now this did produce results in the military.

the Indian race riots broke out in Bombay, and as we left our apartment, British soldiers were setting up a machine gun on our balcony.

We left Bombay, India, with no regrets and no farewells. We had no escort because this was a fast ship. We stopped at Capetown, South Africa, for degaussing. Huge cables were wrapped around the hull and an electric current was applied through the cables. This demagnetized the ship to give it protection from magnetic mines.

During the long voyage home some of our ex-AVG pilots manned the anti-aircraft machine guns. Heavier armaments were provided by two 3 inch .50 cannon in the prow and two 3 inch .50 cannons on the stern. There was also a five inch cannon on the stern. Coast artillery troops manned these weapons. Navy personnel did the communications. I became friendly with some of the gun crew, and they gave me a crash course as a gun pointer. I then spent the rest of the voyage pulling a watch period with one of the gun crew.

Although we had no trouble with enemy submarines, as we drew closer to the States we saw a lot of oil slicks and floating pieces of wreckage. Later we learned that German submarines were having a field day with Allied shipping along the east coast, but not much was being done about it.

Our ship docked in New Jersey—either Port Jervis or Weehawken—the day after Labor Day, September 1942. As I was going through customs, the inspector became agitated when he looked into the RAF first-aid kit I had carried around for months. It contained a morphine capsule with a needle. I had never given it a thought. However, they questioned me thoroughly about it before they let me go, minus the morphine capsule. When I got to the censor official he decided he liked my photos so much he wanted to keep them. He Kept my photos, though he was nice enough to leave me the negatives.

After arriving at my home in Scranton, I registered for the draft. They told me I would not be drafted until January, so I could have a little rest and relaxation. My girl friend, who had never written to me, had gotten married while I was in China. She had somehow forgotten to let me know. In the meantime, I received an "immediate action letter" from the War Department by command of Lieutenant General Arnold. They wanted me back in the Army Air Force. They offered a promotion not less than the grade at which I was discharged. I was told to report to the nearest Air Installation for an interview. Thankfully, my foot locker arrived with all its contents intact.

交通部民用航空局直轄空運隊

CIVIL AIR TRANSPORT

KAI TAK AIRFIELD
HONG KONG

IN REPLY PLEASE QUOTE
OUR REF. NO.............................

CABLE ADDRESS
CLAULT HONGKONG

1 April 1950

Sgt. Charles Baisden
52nd Maintenance Sqdn., Ftr.
McGuire Air Force Base, N. J.

Dear Charley:

I was very glad to receive your letter of March 13th as I had completely lost track of your whereabouts. I am also glad to know that you enjoyed reading my book in which I attempted to tell the story of China from 1937-1945.

It is interesting to know that the off-size machine guns were 7.9 mm. rather than .303. I must have gotten the impression that they were .303's on my visit to the Curtiss factory just prior to the purchase of the Tommyhawks by the Chinese. The Curtiss people told me at the factory that the British were buying the planes less a considerable amount of equipment such as radio sets, gun sights and 30 calibre guns. They were planning to install .303 machine guns because they manufactured that calibre and also had plenty of ammunition for it. You will also recall what a devil of a time we had getting gun sights that would work.

With very best wishes to the wife and three little ones as well as to yourself, I am

Most sincerely yours,

C. L. Chennault

C. L. Chennault
Major General (Retired)
U. S. A.

Letter from General Chennault.

7

Reenlisting, Cadet Pilot to B-17 Gunnery School

In October, 1942, I arrived at the Middletown Air Depot near Harrisburg, Pennsylvania, and was interviewed by a major who had no idea what an aircraft armorer was or what he did. It was a difficult interview, but I did get a Tech Sergeant rating, and in November I reported to the New Cumberland Reception Center. After two weeks of dealing with the fine art of bureaucracy, I received orders to report to Westover Air Field in Massachusetts. Before doing so, however, I had to make some appearances at the base theater where I told them my "tales of heroic deeds and feats of valor."

Upon reporting to the 325th Fighter Group, I was surprised to find three other ex-AVG there, including Ollie Olson. Ollie was Group NCOIC Engineering Inspector, and I was the NCOIC Group Armament Inspector. When we attended the Group Commanders meeting, we were often asked about the AVG. One time the Colonel asked Ollie how he thought the P-47 would stack up against the Japanese Zero fighter. Ollie said, "It's going to make a great aircraft, sir, providing you give it a P-40 escort."

Ollie and I made Master Sergeant on January 1, 1943, but in the meantime, Ollie had talked me into applying for pilot training. I didn't think I met the educational requirements, but I decided to give it a try because the requirements had been drastically reduced. I had no trouble with the test, thus we found ourselves at the Nashville Classification Center in Tennessee, as cadets. This was quite a demotion from Master Sergeant. Officers were allowed to go through the program holding their rank, but enlisted men could not.

During classification testing we were asked to choose between Pilot, bombardier, or navigator training. Ollie and I requested pilot training only. Several other cadets who had done this were eliminated, but I believe our prior service record

Author as an Aviation Cadet-Pilot training—Class 43K-Blythe, Calif, 1943.

Solo flight over Blythe, Calif., in a Ryan PT22 June 1943.

saved us from elimination, and we went through the classification testing without any trouble.

About that time I got sick and had to be hospitalized, and Ollie and I went our separate ways. He was sent to flight school in the south in Class 43J. Later he flew P-47s in Europe. I heard he shot down three enemy aircraft and then was shot down himself. He named his P-47 *Johnny Fauth*.

I was sent to Santa Ana, California, in Class 43K. Here I met the girl who was to become my wife. From Preflight School, I went to Blythe, California, to fly the Ryann PT-22. Also, Willa and I got married here. After completing Primary School, we went to Minter Field, just outside Bakersfield, to fly the Vultee BT-13. My instructor was a captain who hated his job. He didn't spend much time instructing me after I was checked out for solo in the BT-13. I soon discovered I could not do a slow roll or snap roll, and I had a lot of difficulty in understanding how to recover from a spin. Actually, on one occasion, while practicing pylon 8s, I racked the Vultee vibrator into such a tight turn that the aircraft stalled and flipped me upside down and into a spin, which could have been disastrous. When I was thinking of parting company with the aircraft, the BT recovered on its own. This was the first time I realized I wasn't cut out to be a hot shot fighter pilot and that the aircraft was controlling me.

Kingman Army Air Field

Incorporated with "BUGS BUNNY" at KINGMAN, ARIZONA

Hereby Certifies... That M/Sgt. C.N. Daisden has on 11-29th, 1943, Finally Learned What's Cooking About Aerial Gunnery;

So Now

Roger W. Menke

ROGER W. MENKE, 1st Lt., A. C., Commanding.

Shelly B. Waggoner

SHELLY B. WAGGONER, 2nd Lt., A. C., Adjutant.

Bugs Bunny Gunnery certificate from gunnery school at Kingman Arizona, December 1943.

I had no problems with landings, formation flying, or getting from point A to point B, and I was only crashing occasionally when flying the Link trainer. However, I knew I could not fly a combat fighter if I did not have the ability to be master of my own aircraft. This also became obvious to my check pilot.

It was a relief to me to get washed out of flight school, although my pride took a beating. After the war I ran into some of the fellows who had been in my class and they told me many of my class died over Germany.

The elimination board gave me three choices—namely, gunnery, gunnery, or gunnery. With my Master Sergeant's rating back on my sleeves, I proceeded to B-17 aerial gunners school at Kingman, Arizona. Armorers were assigned to either the tail gun position or the ball turret according to height. I was introduced to the Sperry ball, located under the fuselage. It is a fine turret built for a midget. I could not wear a chest chute while inside, and there was some doubt in my mind as to my survivability in the event of a real emergency. Electrically heated suits were issued to us, and these were not too uncomfortable. However, if the electrical power failed, you froze your butt off.

The ground phase training was great fun because I enjoy shooting. We would bang away with shotguns at clay pigeons while driving through the course on a truck. From high towers, clay birds were thrown in the air while we fired on them from our turrets. The turrets were mounted with shotguns in place of the .50 caliber machine gun. We fired the .50s from flexible mounts at targets running by on a track. In a training building, a huge curved movie screen projected fighters making pursuit curves. We fired at these electronically. There was even a sound track, which made quite an impression.

Considerable time was spent in the disassembly and assembly of the .50 caliber aircraft machine gun. The big test was to be blindfolded and then strip the weapon. The whole thing was done entirely by feel. The instructors were prone to add a broken part or two, and the time to pass the test was set at one hour. My best time was twenty minutes, but some of the instructors could do it much faster.

At graduation, I received orders from the Pentagon to report to a base in North Carolina for assignment to Project 9. This caused quite a stir with Kingman because they had planned to keep me there as an instructor and wanted to know who I knew at the Pentagon. I had no idea what was going on and told them so. I don't think they believed me. This was the last class of gunners who received promotions to Staff Sergeants.

I had sent my wife back to her home in Orange, California, shortly after I reported to Kingman because students weren't allowed off base except on weekends, and the housing situation was very bad. I believed I could get some leave time at my new base, so she agreed to go on ahead of me to my home in Pennsylvania. However, I did not see her again for eight months, and this was quite a shock to an eighteen-year-old bride, now 3,000 miles from home, with in-laws she had just met.

When I reported to Goldsboro Staging Center, I found that Project 9 was headed up by my old commanding officer from Mitchell Field, Phil Cochran. He was now a full Colonel. Colonel John Alison was also there, and I had a lot of respect for these two officers. Both of them were outstanding pilots and had a leadership approach that seemed to make any mission possible. Although I had no idea what we would be doing, I knew this was going to be a great organization.

A week after reporting, we went to Seymour Johnson Field and picked up a new C-47 equipped for a glider tow and headed for India via the southern route through South America, Ascension Islands, Africa, Arabia, and into Karachi, India.

8

The First Air Commandos

The First Air Commando Force was the result of a conference held in Canada by top allied brass. British Admiral Lord Louis Mountbatten and General Arnold discussed a highly mobile fighting unit for the CBI theater to support the Chindits under General Wingate. General Arnold's conception of this unit was one complete with all supporting services. As the unit evolved it would change names four times: (1) Project 9; (2) Project CA281 (Rumor had it that was the hotel room number where Colonels Cochran and Alison met and planned their strategies); (3) 5318th Provisional Unit (Air); and (4) No.1 Air Commando Force. It became known as the 1st Air Commando Group in 1983 through our veterans organization.

I arrived in Karachi, India, in late December 1943, and was met there by Major R.T. "Tadpole" Smith, ex-AVG pilot from my old 3rd squadron, who had a devious sense of humor. He was the one who had submitted my name to the Pentagon. He greeted me with a grin and said, "So you didn't like Kingman?" Then I remembered while at gunnery school I had written Major Smith when he was squadron commander of a P-38 outfit at Van Nuys. I hadn't liked Kingman and told Smith so.

After spending Christmas in an Indian quarantine station hospital because my yellow fever immunization shots were out of date I reported to my unit. It was the 5318th Provisional Unit (Air). I had never seen so many NCOs. There were more staff, tech, and master sergeants than any other enlisted grades. There were P-51 fighter, C-47 cargo, glider tow L-1 and L-5 light planes, YR-4 helicopter, UC-64 utility, and CG4A glider pilots, technicians, and support personnel all over the place. I realized with Colonels Cochran and Alison, Majors R.T. Smith, Grant Mahoney, and Avrid Olson as our leaders, whatever we were going to do was going to be done and done right. These were men I would follow "anywhere, anytime, and anyplace."

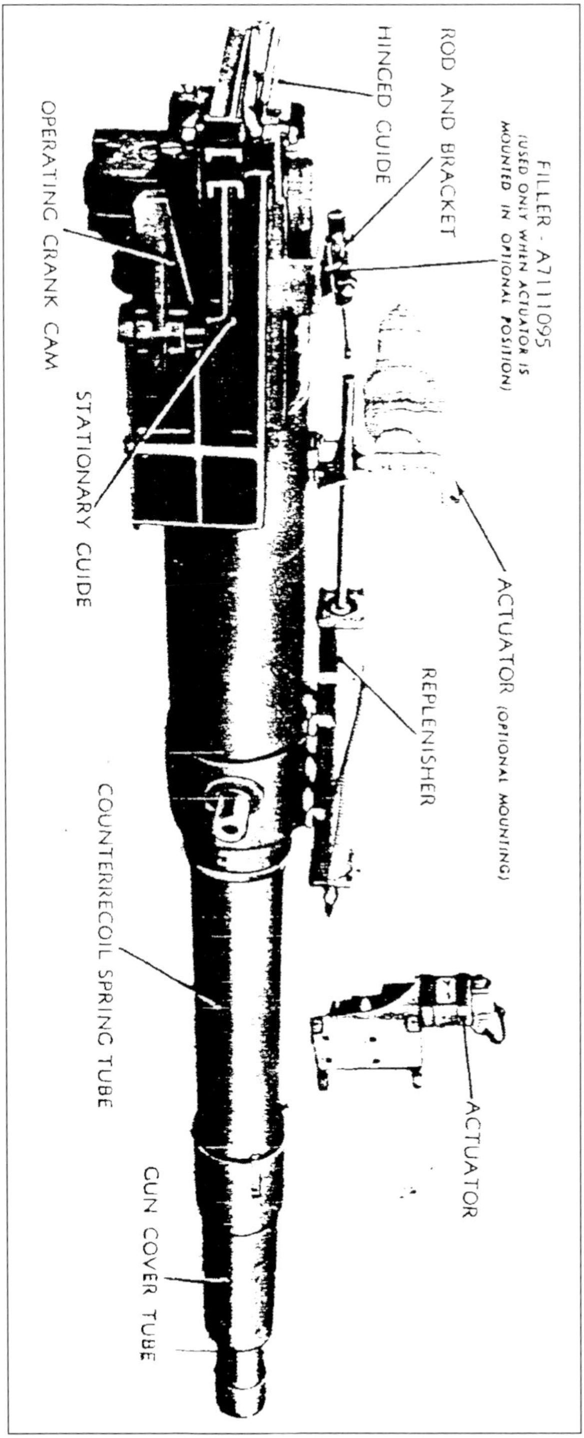

75mm AN-M5 Aircraft cannon—one of 2 types used in the B-25H.

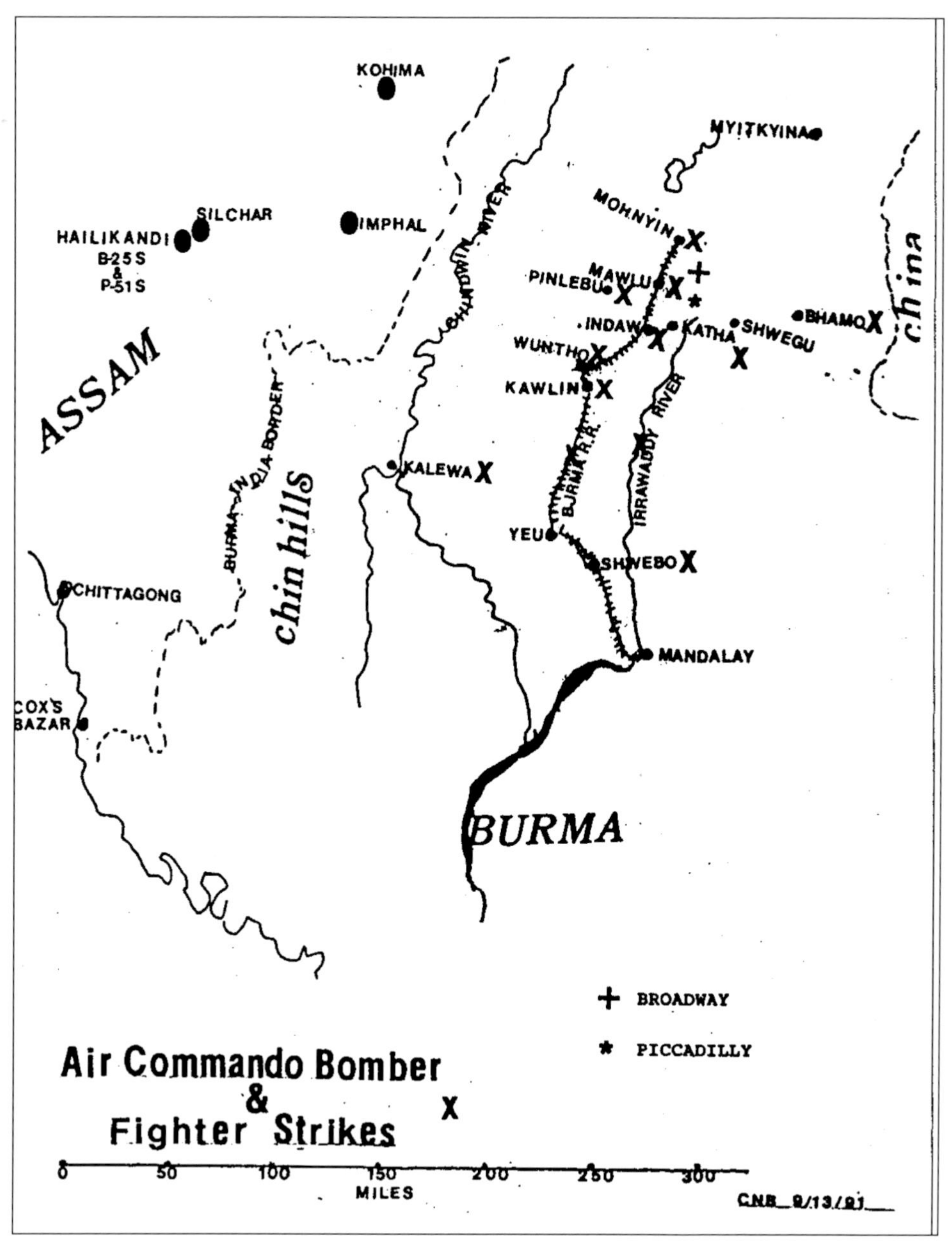

Map of Air Commando Operations in support of the Chindits.

"Dive Inn" No cover Charge Chuck Baisden, prop. Hailakandi, Assam, India, 1944.

The First Air Commandos was the finest organization I was ever associated with during my service career. You were told what the mission was, and then, depending on your job, you proceeded to do it without the fuss and bother of someone looking over your shoulder and telling you how to do it. This was very similar to the AVG, but here we had some military discipline.

Shortly after I started working in the fighter armament section, a B-25 bomber section was added to the group. These consisted of eleven B-25H and one B-25J models. The Hs had the .75mm cannon and the J model had the glassed-in nose for a bombardier. R.T. was made commanding officer since he had twin engine P-38 fighter experience, and he was promoted to Lt. Colonel. I immediately accepted his offer to be his turret gunner and NCOIC of the bomber gunners.

Since the bombers were not part of the original plans, and now we had twelve bombers, each requiring a crew of five in the Hs and seven in the J model, R.T. took off to other bomber bases to find crews for these aircraft.

Before the new crews arrived, I bore sighted several of the bombers and made myself familiar with the .75mm cannon, the turrets, and the bombing systems. When R.T. returned, we test fired our weapon systems on a range west of Karachi (Sonmiani Bay). The range was a small rocky group of little islands. We fired our weapons over the water onto the rocks, and, on some occasions seagulls had a number done on them.

On these test flights there were only R.T., Lt. Weber, our navigator, and myself. R.T. fired the .75mm cannon, as well as the nose and side package guns. I would load the cannon, fire the upper turret, and then go to the back and fire the waist guns and tail turret. Until the rest of the crew arrived, I was kept busy cleaning guns and loading ammunition, but I was as happy as a kid with a new toy.

Around the first of February we left Karachi and flew to our operational base at Hailakandi, Assam. Hailakandi is located in the tea-growing region of Assam, eighty miles west of Imphal and one hundred and twenty miles from the Burma border. The runway was not paved and was suitable for fair weather operations only. The bombers were dispersed in the surrounding jungle some distance from the runway. In our particular instance they were located next to a villager's thatched hut and a small pond. The villagers had already been evacuated from this area. It was about a five minute walk to our operations hut (basha) from our dispersal area, which was hidden in the jungle. We had thatched huts with folding cots and a mosquito net, our only comforts. A single low-wattage bulb made it a dark, uninviting place;

Washing clothes at our bomber dispersal area at Hailakandi.

somewhere you didn't want to spend much time. Hot showers were possible if you got there first, because the sun heated the galvanized pipes that ran for some distance from the water source. We traveled back and forth from our living quarters to the aircraft by truck.

I devoted almost all my time to our aircraft and didn't have much to do with the supervision of the gunners. They knew their jobs and their responsibilities and what might happen if they screwed up.

The mess hall provided a monotonous diet of fried, boiled, and baked spam, dehydrated eggs, and potatoes. Instead of bread we had hard crackers. We also had to contend with the awful after taste of the vitamin pills which were shoved down our throats while we stood in the chow line. There was a small portable gasoline stove that came with the aircraft's emergency equipment, so I tried cooking a pot of dehydrated beans once in my GI helmet. The beans burned, and my helmet had a peculiar odor left in it which I noticed every time I had an occasion to wear it. I also bought chickens and ducks from the local natives with the thought of trying to cook them, but a mongoose got the chickens before I did.

One of the best boxed rations that came out of WWII was the 10-in-1. The ration would supposedly feed ten men for one day or one man for ten days. Among the best items were Hormel's canned bacon, rice pudding, and baked beans. These rations were supplied to all the bomber aircraft and were so popular that we were threatened with a court martial for opening them unless it was a real emergency. Well, we had a lot of emergencies!

Although I did gripe about our food, I didn't realize what bad food was until I ate in a British enlisted mess. During our flights into Imphal, we swapped spam and vienna sausages for their corned beef and marmalade. They were as tired of their rations as we were of ours.

Our armament officer was Captain Andrew Postlewait, who, along with his crew of ordinance specialists, kept our aircraft supplied with many types of munitions. The bomber crews loaded their own planes and took care of their own weapons. It was not unusual to find one hundred, five hundred, or one thousand pound general purpose bombs, crates of fragmentation cluster bombs, bundles of incendiary clusters, boxes of .50 caliber machine gun ammunition, and stacks of tubes containing .75mm rounds stacked around each bomber's dispersal area.

A typical combat mission day would usually start with the enlisted crew members arriving at their aircraft to do a preflight check while the officers were being briefed. We were briefed at the aircraft since the briefing room was too small for everyone. I never did know how our radio operators knew what was going on, and what crystals they needed to use for their transmitters.

The bomber crews also loaded up their aircraft the day before the mission, or as soon as they landed from a previous mission. There were a few times we would have to download and reload if a different type of load was required. That would mean downloading six five hundred pound general purpose bombs and reloading with twelve fragmentation cluster bombs, or twelve incendiary cluster bombs. Once the aircraft was properly loaded, everyone would climb aboard. I would have time for one cigarette, then it would be time for me to get into my bicycle-type seat in the Bendix turret and get permission to fire the machine guns.

Our Air Commando bombers did not meet any Japanese fighters. In fact, I met only one Jap fighter, and that was on my last mission. Our gunners did strafe ground targets and caused many Japanese casualties. However, it was very difficult for a radio operator to hit much with his waist guns. He could only use one of the two waist guns, and we were so low and went by so fast that he did not have much opportunity to fire at ground targets. If he was firing during pull up from a target, it was very easy to shoot holes in the vertical stabilizers as the centrifugal force pulled him down and the muzzle went up. The gunners did very little firing when we were working close support with the ground forces because of the possibility of rounds going into our own lines. This did happen once when a tail gunner shot a Chindit and some mules.

There were no fire interrupters on the top turret, but there were steel plates over the top of the fuselage, directly behind the turret. They looked like two bumps. They were there to prevent the turret gunner from blowing off the head of the tail gunner. It was possible to swing the turret forward and join in with the pilot when he was firing the nose weapons. However, this action placed the muzzle of the two noisy .50 machine guns just a few inches to the rear of the pilot's head, and the result was usually a very unhappy pilot.

Much has been written about the .75mm cannon. In my opinion, it was a fine weapon. It was very effective against such targets as trains, barges, and trucks. It had an explosive effect when fired at longer ranges, and this caused the opposition to keep his head in the dirt. It did have the drawback of being a single shot weapon, and it was only as accurate as the pilot who fired it. Fortunately, I had a pilot that could shoot it like a rifle. Incidentally, contrary to another story, I never fired the cannon. To do this I would have had to sit on the pilot's lap, and I believe he would have strenuously objected. The .75 was fired by electrical solenoid, with the firing button located on the pilot's control yoke.

I loaded this weapon many times during some of our missions. I found we could get off three rounds under combat conditions. Normal recoil was twenty-one and one-half inches. During firing, the counter recoil opens the breech and closes and cocks automatically when the next round is inserted. Muzzle velocity of the high explosive shell was 1,970 feet per second, which got to the target some time before we did. There was a jolt when the weapon was fired, but it was not excessive, and anyone who says the aircraft stopped in midair has never flown in the aircraft during firing. There was a small spring-loaded trap door just behind the gun breech where empty shell casings could be dropped out through the bottom of the fuselage. If I was in the turret and facing aft during firing of the cannon, it was wise for me to have my feet firmly on the turret gun charging pedals, because the cannon recoiled uncomfortably close to my foot.

The four nose guns were charged prior to take-off because there were no gun chargers in the cockpit. The two package guns on the right side of the fuselage could be charged in flight.

Our aircraft did not have dual pilot controls. Instead, a jump seat was installed for a navigator or observer on the right side. No Norden bomb sight was carried. Instead, we had a modified optical gunsight. A fixed ring and bead sight was attached to the outside of the fuselage in front of the pilot's wind screen. Armor plate was provided on the back of the pilot's seat and on the left side of the fuselage. There was also armor plate in the top turret and in the tail gunner's position.

9

Journal Entries

Combat Mission No. 1
February 12, 1944, 3:00 Flight Time [Note: All times are actual flying hours]
From Hailakandi, Assam, to British Hdqtrs, Imphal. Picked up General Orde Wingate and flew reconnaissance mission to Katha, Burma. Destroyed small railroad bridge and blew the roof off a large building with .75mm shell fire. Small arms fire from ground resulted in rifle bullet in fuselage lodging in .50 caliber ammunition chute and just missing tail gunner, Sgt. Miller.

Combat Mission No. 2
February 13, 1944, 2:14 Flight Time
From Imphal to Meza, Burma. Reconnaissance and photo mission of landing fields. British observer with us was one of Gen. Wingate's officers who had walked out of Burma last year. Attempted to destroy Jap bridge with bomb and .75mm shell fire. Results were lacking. A five hundred pound bomb skipped over a bridge and blew up in the jungle 100 yards away. The shell fire did no good because it would pierce the bridge and explode in the water.

Combat Mission No. 3
February 14, 1944, 2:45 Flight Time
From Imphal to landing strips in northern Burma. Reconnaissance and photo mission.

Combat Missions No. 4 through 12

February 17-18, 21-23, 25-26, 28-29, 1944, 3:00 Average Flight Time

From Hailakandi to various places in northern Burma. Reconnaissance and photo primary missions with targets of opportunities our secondary mission. Used five hundred pound G.P. (general purpose) bombs with eight to eleven second delay fusing. Destroyed several bridges and railroad lines. Skip bombing from altitudes 50, 75, 100, and 150 feet at 245 miles per hour was attempted, but the bombs had a tendency to ricochet from the target area. We then resorted to lobbing them in from around 200 to 300 feet. This method proved very successful. No enemy air activity was encountered on any of these missions, although we knew the Japs had over 300 various types of combat aircraft in the Burma-Thailand theater. Ground fire from small arms was always in abundance, but very inaccurate. We also strafed many Jap villages and dispersal areas.

Combat Mission No. 13

March 1, 1944, 3:15 Flight Time

No details of results noted.

Combat Mission No. 14

March 2, 1944, 3:35 Flight Time

From Hailakandi to Pintha, Burma. We destroyed several engines with .75 mm and .50 caliber machine gun fire. Severe damage was done to rolling stock, but because the Japs always move their supplies at night, the cars were empty. Colonel Smith placed a .75mm armor piercing shell into the boiler of one engine and the steam squirted up to a height of 200 feet. I managed to get some nice bursts from the upper turret and had the satisfaction of seeing my incendiary bullets explode on the engine.

Combat Mission No. 15

March 3, 1944, 3:10 Flight Time

From Hailakandi to Kyaithin, Burma. Attempted to bomb storage depots and dumps. Fairly good results, but no definite damage could be noted. Fires started in several buildings.

Combat Mission No. 16

March 3, 1944, 3:15 Flight Time (night)

Night formation over Japanese lines. Dropped one thousand pound G.P.s at intervals. Object was to get their attention and make them a bit jittery.

Combat Mission No. 17

March 4, 1944, 3:05 Flight Time

Hailakandi to Lonkin, Burma. Photo mission of landing strip. [From my memory of this occasion, this was the field covered with logs. At that time it was thought they were placed there by the Japanese to prevent landings. The results of this mission caused much concern with the brass. There was some talk that the mission had been compromised. The force going into this field was diverted to another field. Later on it was determined that the Burmese had placed the logs there to dry].

Combat Mission No. 18

March 5, 1944, 3:05 Flight Time

Hailakandi to Okkyi, Burma. Photo mission of landing strip. This field was one not covered with logs. Our gliders went in during the night and established a fairly good runway.

Combat Mission No. 19

March 6, 1944, 2:30 Flight Time

Hailakandi to Inywa, Burma. Bombed warehouses and oil storage depot using incendiary and fragmentation cluster bombs. Fifteen fighters (P-51s) escorted nine bombers. The fighters were loaded with two five hundred pound bombs and six 4.5" bazooka-type rockets. Huge explosions and fires started.

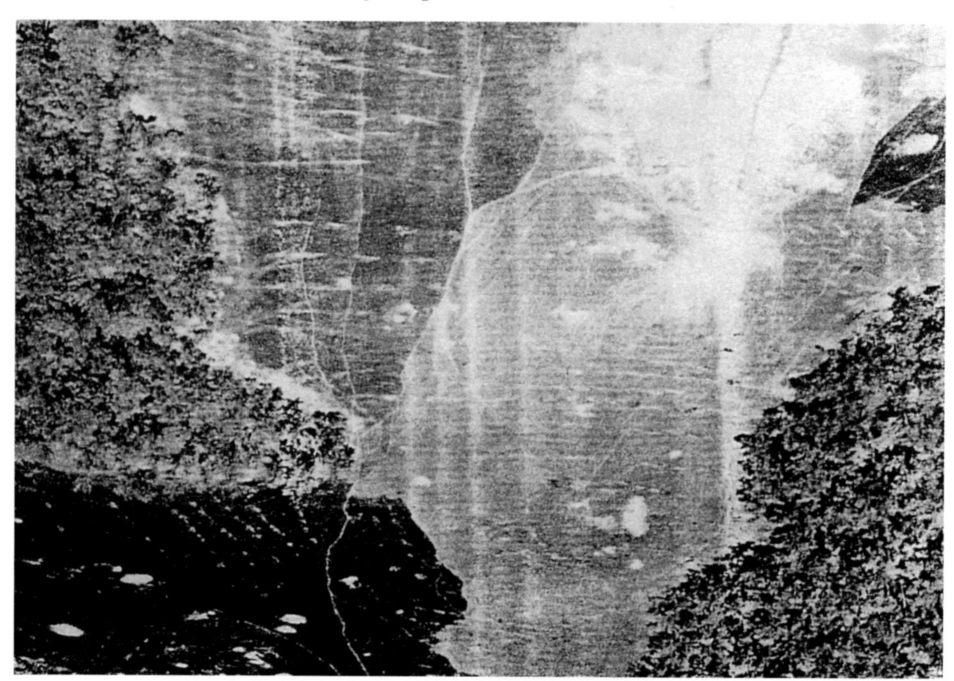

The landing site code named Piccadilly, *referred to in mission no.17.*

Combat Mission No. 20

March 7, 1944, 3:40 Flight Time

Hailakandi to Bhamo, Burma. Twelve B-25s dropped one hundred and five hundred pound bombs on the Bhamo Air Field. We used six to twelve-hour chemical delay fusing in some of the bombs. The runways and taxi strips were thoroughly plastered, and the field was definitely of no use for the Japs to attack "Broadway," code name for landing site in Burma.

Combat Mission No. 21

March 8, 1944, No Flight Time Given

Hailakandi to Katha, Burma. Bombed railroad and rolling stock, warehouses, dumps, and destroyed radio station.

Combat Mission No. 22

March 8, 1944, 3:30 Flight Time (night)

Hailakandi to Shewbo, Burma. Night mission on Shewbo Air Field. Dropped incendiary and fragmentation cluster bombs. Destroyed twelve Jap aircraft, blew up fuel dumps. Left huge fires burning. Encountered anti-aircraft fire from heavy guns in the town of Shewbo. Japs did not disclose their position until after we had bombed the field. No search lights were observed, and, though their fire was heavy, it was not very accurate. On this raid Lt. Weber, our navigator, won himself a spot in all our hearts. Though it was a moonless night and there was the usual haze from the countless forest fires, Lt. Weber brought us directly to the Jap air base and then back to our own base.

Combat Mission No. 23

March 11, 1944, 3:00 Flight Time

Hailakandi to Katha, Burma. Bombed dumps and supply areas with good results.

Combat Mission No. 24

March 11, 1944, 5:00 Flight Time (night)

Hailakandi to Heho Airdrome, Burma. Mission flown at night, which proved unsuccessful. We had a lot of difficulty locating the target and would not have found it if the Japs hadn't turned on their search lights. Only one bomber released bombs, and they exploded a mile from the field. We almost had a mid-air collision when the search lights bracketed the lead formation, blinding the pilots momentarily. (Being hit with search lights at night while flying formation is not exactly habit-forming.)

Barbie III leads a gunship formation on a mission over Burma. March 1944. AF Photo

Combat Mission No. 25
March 12, 1944, 3:00 Flight Time
Hailakandi to Dineblu, Burma. No results observed.

Combat Mission No. 26
March 13, 1944, 2:55 Flight Time
Hailakandi to Wuntho, Burma. The Chindits had established a roadblock at Wuntho, and through radio directions and targeting the area with colored smoke, we were able to plaster the Jap position.

Combat Mission No. 27
March 14, 1944, 2:35 Flight Time
Same target and mission as the day before. Damage was very hard to see because of jungle undergrowth. British reports said the bombing was very effective.

Combat Mission No. 28
March 14, 2:40 Flight Time
Hailakandi to Meza, Burma. Bombed Meza railroad bridges with excellent results. Two direct hits on the bridge and one on tracks just short of the bridge. The fighters dive-bombed a truck pontoon bridge a few hundred yards north. Both bridges were destroyed.

Air Commando gunships over Burma. AF Photo

Combat Mission No. 29
March 15, 1944, 2:35 Flight Time
Hailakandi to Kawlin, Burma. Bombed supply dumps. No results noted.

Combat Mission No. 30
March 16, 1944, 2:45 Flight Time
Hailakandi to Kalu, Burma. Results unobtainable.

Combat Mission No. 31
March 17, 1944, 2:35 Flight Time
Hailakandi to Malu roadblock, Burma. Bombed Japanese positions.

Combat Mission No. 32
March 18, 1944, 2:45 Flight Time
Same mission as yesterday.

R.T. (Roundtrip Smith) and crew of BARBIE III. AF Photo

Air Commando B-25 crews. Hailakandi, Assam, India, 1944. AF Photo

Air Commando B-25 Pilots and Navigators AF Photo

1. Capt. Randolph E. Green
2. 1/Lt. Brian Hodges
3. 1/Lt. Archie McKay
4. 1/Lt. Emory Taylor
5. 1/Lt. Murrell Dillard
6. Capt. Carl Ziegler
7. 1/Lt. William Burns
8. 1/Lt. Ralph Lanning
9. 1/Lt. Stephen Wanderer
10. Capt. Frank Merchant
11. Capt. Daniel A. Sinskie
12. Unknown
13. 1/Lt. Frank Gursansky
14. Unknown
15. 1/Lt. Wesley Weber
16. 1/Lt. Leslie Nielson
17. Major Walter Radavich
18. Lt./Col. Avrid Olson, Operations Officer
19. Colonel John Alison, Co. Commander 1st ACG
20. Colonel Philip Cochran, Co. Commander 1st ACG
21. Lt./Col. Robert T. Smith, Commander B-25 Section

Combat Mission No. 33

March 20, 1944, 2:55 Flight Time

Hailakandi to Indaw Lake, Burma. Reached objective around 1600 hours. Caught fifteen Jap trucks loaded with troops on a road just west of town. Road was in a narrow defile. Blocked first and last trucks in convoy with .75mm shell fire and went into our gunnery/bombing pattern. Completely destroyed convoy with machine gun cannon and frag bombs. I got good bursts into the trucks as we pulled out from our strafing runs. Caused nearly 300 casualties and brought a thirty-six-hour lull in Japanese resistance. Also caught five locomotives. Damaged two and blew the boilers out of three. I used our entire amount of .75mm shells (21), plus a few I had stashed away on the cockpit floor.

Air Commando P51s escorting a B-25 formation over Burma 1944. AF Photo

The 1st Air Commando flew North American P51-A Mustangs.

Combat Mission No. 34
March 21, 1944, 2:50 Flight Time
Hailakandi to Meza, Burma. Bombed Meza bridge. It was successful.

Combat Mission No. 35
March 22, 1944, 3:10 Flight Time
Hailakandi to Malu, Burma. Bombed Jap troops at roadblock.

Combat Mission No. 36
March 23, 1944, 3:15 Flight Time
Hailakandi to Indaw, Burma. Bombed Japanese supply dumps. Bad weather forced us to land at Broadway. Had to roll 55-gallon drums through dense elephant grass to our aircraft and refuel by a hand pump. Spent the night on plane.

Combat Mission No. 37
March 24, 1944, 1:40 Flight Time
Broadway to Hailakandi, Assam. Took off at sunrise and flew directly to home base. Jap aircraft attacked Broadway minutes after we left area. Broadway is one hundred and fifty miles into enemy territory.

Combat Mission No. 38
March 27, 1944, 2:45 Flight Time
Hailakandi to Meza, Burma. Bombed and destroyed bridge. The Japs repair this bridge as fast as we blow it up.

Combat Mission No. 39
March 28, 1944, 3:00 Flight Time
Hailakandi to Mahnyin, Burma. Bombed Japanese troop position. I was very sick prior to and during the mission. I had chills, fever, and a terrific backache.

(March 29 through April 7, 1944, I was in the hospital at Hailakandi. I had malaria, evidently from our overnight stay at Broadway. I got the nine-day quinine cure and a short pass to Calcutta. While I was in the hospital, the GI in bed next to me with malaria and mumps died. I did not even know his name.)

Combat Mission No. 40
April 8, 1944, 2:50 Flight Time
Hailakandi to Malu, Burma. Bombed Japanese troop area.

Combat Mission No. 41
April 9, 1944, 2:55 Flight Time
Hailakandi to Indaw, Burma. Bombed Japanese supply dumps.

Combat Mission No. 42
april 10, 1944, 3:00 Flight Time
Hailakandi to Banmauk, Burma. Bombed Japanese supply dumps.

Combat Mission No. 43
April 11, 1944, 2:30 Flight Time
Hailakandi to Naumgkan, Burma. Bombed ground troops and strafed village.

Combat Mission No. 44
April 16, 1944, 3:00 Flight Time
Hailakandi to Mahnyin, Burma. Bombed supply dump.

Combat Mission No. 45
April 19, 1944, 4:05 Flight Time
Hailakandi to Mawlu, Burma. Bombed Japanese troops.

Air Commandos Lt/Colonel Avrid Olson and Colonel Phillip Cochran confer with Chindit Leader, Orde C. Wingate. AF Photo

Combat Mission No. 46
April 19, 1944, 2:55 Flight Time
Same mission as No. 45

Combat Mission No. 47
April 21, 1944, 2:55 Flight Time
Same mission as No. 45-46.

Combat Mission No. 48
April 23, 1944, 3:05 Flight Time
Hailakandi to Indaw, Burma. Bombed Japanese troop concentration and strafed village. Last mission with Lt.Col. R.T. Smith. (An expanded version of bombing the village follows.)

At 1300 hours on April 23, 1944, eight B-25s escorted by ten P-51 fighters took off from Hailakandi on what was one of the most ruthless, yet necessary, raids I had ever been on. Our objective was a Burmese village on the north shore of Indaw Lake. A Chindit column was advancing from the main stronghold, the roadblock at Mawlu. They needed the water here for drinking.

British Admiral Lord Louis Mountbatten, Supreme Commander South-east Asia Command, with Colonel Phillip G. Cochran and Lt. Colonel Robert T. Smith, Hailakandi, Assam, 1944. AF Photo

We arrived over our target at 1500. The weather was fair, but there was poor visibility because of the haze from burning forest fires abundant at this time of year throughout Burma. Our aircraft, "Barbie III," led the first flight of three over the target and released our fragmentation cluster bombs (para-frags) at an altitude of about two hundred feet. Our bomb cover was good, and fires immediately broke out in several parts of the village.

Captain Sinskie led his formation over next, but he released his incendiary clusters too late, and they overshot by some two hundred yards. Captain Zeigler came over last and dropped his para-frags. They covered the entire town and fires immediately came into view. Smoke rose to two thousand feet. The fighters came in last and dive-bombed a small group of huts about one-half mile northwest of our target. They were carrying two five hundred pound GP bombs per aircraft.

As soon as the fighters finished and got altitude above us, R.T. ordered two flights into our gunnery pattern, and we commenced strafing the village. Our .75mm shells tore into the thatched buildings and literally exploded them into a thousand pieces. I would hear the heavy rattle of our six .50 caliber machine guns and the dull thud and jolt of our cannon, and a thousand yards away a hut (basha) would explode in a red flash and black smoke. I could see our tracers hitting the ground and ricocheting into the air. They appeared to eat their way into a basha. We averaged three bursts of machine gun fire and three cannon shells per run.

After each aircraft had made two passes, we concentrated our fire on the bashas the fighters had dive-bombed. Two runs were made on these, and then we started working over the flat bushy stretch of ground starting at the lake edge and then into the town. This area contained trenches and dugouts which I strafed from my turret as we pulled up from a run. I couldn't see the results from my fire.

The next day we bombed Indaw, and we circled over the lake where we had been the day before, I looked at what was left of the Burmese village. There were just dozens of spots of gray. The town was just not there. I wondered what their casualties were. I wondered what damage they suffered and what they thought.

I flew the following missions as a spare gunner. I flew with such pilots as Captain Ziegler, Major Radovich, and others whose names I don't remember. I usually flew as a tail gunner or turret gunner depending on what position I was needed to fill.

Combat Mission No. 49
April 23, 1944, 3:15 Flight Time
Hailakandi to Indaw, Burma. Bombed Japanese supply dumps and railroad.

Combat Mission No. 50
April 24, 1944, 3:05 Flight Time
Hailakandi to Hopin, Burma. Bombed Japanese supply dumps.

Combat Mission No. 5l
April 27, 1944, 3:10 Flight Time
Hailakandi to Meza, Burma. Did not bomb due to weather.

Combat Mission No. 52
May 5, 1944, 3:20 Flight Time
Hailakandi to Hopin, Burma. Bombed village that contained Japanese troops and supplies.

Combat Mission No. 53
May 5, 1944, 3:00 Flight Time
Went back and did it again.

Combat Mission No. 54
May 6, 1944, 3:55 Flight Time
Hailakandi to Ritpum, Burma. Bombed village that contained Jap troops. On this mission we were carrying twelve frag cluster bombs (three bombs per cluster). Minimum altitude for releasing these bombs was around 2,000 feet. We dropped these going over the target not much over 200 feet. When we landed at Hailakandi, we found the bottom of the fuselage was riddled with holes. Sgt. Zajak, who was

No.1 USAF Air Commando leaders—Colonels Phillip G. Cochran and John R. Alison, Hailakandi, Assam, 1944. AF Photo

Japanese propaganda dropped on British troops in Burma, 1944.

Copy of Allied propaganda dropped on Japanese troops in Burma, 1944.

flying spare waist gunner, counted fifty-nine holes. The pilot claimed it was from ground fire, but we knew he had just screwed up. I think he thought we were carrying para-frags.

Combat Mission No. 55
May 6, 1944, 3:00 Flight Time
Hailakandi to Rinbaw, Burma. Bombed village that contained Jap troops and supplies.

Combat Mission No. 56
May 11, 1944, 3:25 Flight Time
Hailakandi to Nalong, Burma. Bombed Japanese troops.

Combat Mission No. 57
May 15, 1944, Flight Time Not Logged
Hailakandi to Silhet, Assam to Namkwin, Burma. Bombed Japanese troops.

Another example of Allied propaganda.

Barbie III after 40 missions.

The crew of BARBIE III. AF Photo

Combat Missions No. 58

May 16, 1944, Flight Time Not Logged

Hailakandi to Silhet, Assam, to Hopin, Burma. Attempted to destroy C-47 that had crash landed with classified equipment aboard. We arrived over target to find both Japanese and Chindit patrols engaged in the target area. A Japanese fighter made a pass at us from six o'clock level, but he broke off to three o'clock when I fired a burst at him from the tail turret. Both guns jammed after the first burst. I got no response from the front end, unaware that my mike was inoperative, and could see the radio operator was asleep. I woke him up by throwing .45 cartridges at him. I had taken them from my pistol clip. The crew did not believe me when I first reported this upon our return to base. However, the tail gunner on our right confirmed my sighting. The Jap fighter had us, cold turkey, but I believe he broke away because he feared our P-51 escorts. However, we had no escorts on this mission! In the end, we did not attempt to bomb the crashed C-47.

I do not know if other gunners had problems with the Bell hydraulic tail turret as I did, but every one I test fired had one kind of problem or another. The .50 caliber metallic belting links would tend to jam together in the common link discharge tube, a stoppage very difficult to clear in the air. The ammunition containers were several feet from the weapons and required booster motors to force the ammunition to the guns (these burnt out constantly). When the air temperature rose inside the aircraft under a hot sun, the hydraulic fluid would overflow from the reservoir, making a slippery mess in the gunner's compartment. A manual turret like the B-17 had would have been a big improvement.

10

Barbie II, Barbie III, and Other Details

Our aircraft was named after our pilot's wife, Barbara, who married R.T. in June, 1943. *Barbie* I was a P-51 that R.T. flew when first assigned to the Air Commandos. *Barbie* II was our first B-25H assigned to our crew in Karachi and flown by us to our forward operational base at Hailakandi, Assam.

During the months of March and April, 1944, we were involved in numerous strikes on the Burmese railways. It was during one of these missions, while firing our .75mm cannon and nose machine guns at a camouflaged train, that we lost the nose gunnery access hatch due to the locking latches coming loose from the recoil vibrations of the firing. When the hatch came off, it shattered the pilot's wind screen, and when we landed, a five hundred pound bomb that we had not dropped on the mission fell out through the bomb bay door and ripped through the bottom of the fuselage. A photographer who was in the aft section was thrown around and injured.

While on this mission, we had a guest named Dick Rossi, a civilian pilot for the China National Airways. Dick had also been a pilot in the AVG and a good friend of R.T. and myself. After this mission, Dick didn't ask for a second ride.

Barbie III, tail number 34280, was flown by our crew until April 23, 1944, when Lt.Col. R.T. Smith was ordered back to the States. The crew continued to fly with other pilots.

Mission Incidents

On one occasion, Colonel Cochran flew with us. He was sitting under my turret and when I started to bang away at a target, the hot .50 caliber shell casings landed on his bare arms. He looked up at me and yelled, "Keep shooting!," and then got out of the way.

On March 4, 1944, Lt. Murrell Dillard and crew were lost on a mission. I saw them crash, but I have no record that we were on this mission. We flew a photo mission late in the afternoon on that day, and I must have flown one earlier in the morning. I distinctly remember Major Radovich dropping his one thousand pound bombs on a barge about the time Dillard's plane went in.

Tech Sgt William Postlewait, one of the gunners, was a good friend of mine and a brother of Captain Postlewait, our armament officer. He came up to me after we returned from our mission and asked where his brother's plane was. I had to tell him that it had gone down. He never said a word, just walked away.

Birth of My First Gray Hairs

I was working alone at our dispersal area uncrating para-frags. These are twenty pound fragmentation bombs that come in a cluster of three. When dropped, these bombs deploy a parachute from the rear and allow for dropping at a very low altitude. The bombs are armed through a fuse delay, which ignites upon the opening of the parachute. As I lifted a cluster from the packing crate, a nail caught on the nylon arming line and pulled the safety pin from the nose of the fuse, arming the bomb.

As I stood there holding a high explosive bomb with a smoking, sizzling fuse, I realized I had really screwed up and probably wasn't going to live much longer. In a few seconds, the fuse was quiet, so I very gently laid the bomb cluster on the ground and left the area. After a few moments to recover from the shakes, I found one of our ordinance people, who removed the fuse and threw it in a little pond next to our aircraft. It exploded with a very large bang. He told me if I had dropped the bomb, it would have blown up the entire area. It would have only taken four and one-half pounds of force to set it off!

Another hairy incident involved the .50s on the right side of the fuselage. They were charged after take-off by using a small hand-charging device that looked like a small cargo hook. One time I was charging one of these guns when my hook caught the right engine CO2 extinguisher cable. I accidentally discharged the CO2, much to the surprise of R.T. and Lt. Weber.

One time a frag cluster got hung up in the bomb shackle and failed to release as we dropped our load, so I got into the bomb bay through the hatch hole above the bomb bay. By hanging on one of the shackles with one hand and placing my feet on the bottom of the bomb rack, I released the cluster manually, letting the cluster slide down between my body and the bomb rack, then letting the whole thing slide down the bomb bay door. R.T. then opened the bomb bay and the cluster fell out.

On the mission where we lost our nose gunnery hatch, I was looking out of my turret after we had landed and saw the bomb bouncing around in a huge cloud of dirt and dust on the runway. Two guys in a Jeep began chasing after this unknown object. It was comical to see their reaction when they realized what it was they were chasing! No one has ever done a one hundred and eighty degree turn in a Jeep any faster than those guys. Actually, there was little danger of the bomb exploding, because it had not dropped far enough for the arming vanes to have spun off. But they didn't know that.

We never used our oxygen, and I don't ever recall wearing an oxygen mask. Some of the bomber gunners who came in from other outfits wore chest pack parachutes, but our crew only had the seat type chutes. It was impossible to wear this type of chute in the upper turret. Most of our flying was done at low altitude, and they were of no use anyway.

We were flying a reconnaissance and photo mission with one of the photographers who insisted on firing one of the waist guns. Sgt Dixon, our radio operator, reluctantly let him try, and he promptly shot a hole in our left vertical stabilizer.

During another bombing and strafing mission, as we pulled up from our run and were circling around for another pass, I saw a large herd of very frightened water buffalo stampeding across the open field toward the town we were bombing. They ran right into our descending parachute frag bombs. This was not a pretty sight.

There was a brief period of time in which we were not being briefed prior to a mission. This was not well received by our enlisted crew. We knew there were OSS and Chindit patrols in the areas we were flying over, and we wanted to know which direction to head if we had to make a forced landing. We all realized, though, that our chances of survival were not good should we go down near one of the towns we had bombed. After our protests, the briefings were started again.

11

Summary of Air Commando Operations

When the rainy season began in late May of 1944, our operational airfield at Hailakandi was abandoned. The runway turned into a mud hole, so our bomber and fighter support missions for the Chindits came to a halt. We moved to our new base at Asonsol, India. Our pace of work slowed down considerably, and no one seemed to know what was going to happen next.

The following is a copy of a letter posted on our bulletin board there. I have no record of who wrote this letter or to whom it was written:

Dear General:

I am glad to inform you that the First Air Command Force has been a complete success in accomplishing its first mission. As Col. Alison said when he returned from "Broadway," "They told us to delay the Japs in getting into the area, and we delayed them. They told us to set up air bases behind the Jap lines and put thousands of troops and equipment into Burma in gliders and transports, and we did it. And they told us to protect them from air attack after we got them, so we wiped out everything they sent up. They didn't think we'd do it, but we did it, and what's more, we did it on time."

The past six days have been busy ones. In fact, no one has had much time for rest since operation "Thursday" began last Sunday afternoon. The beginning was overcast, with dark premonitions of failure. The briefing of officers and air crews had just been completed when the photo office came in with prints of reconnaissance photographs taken that afternoon. These showed the middle on one of the three fields chosen for the landings had been systematically strewn with freshly cut trees. The trees had been lain in such a manner that it would have been impossible to land any type aircraft on it.

Wingate, Cochran, and Alison were then quite certain the enemy was aware of our intention. "Broadway," the northern strip, was clear, and after a good deal of deliberation, it was decided to send both the Broadway and Piccadilly force to one field.

The first tow of two gliders carrying Col. Alison and the radio and field equipment took off at 1840, seventy minutes behind schedule. These two and the two operations gliders destined for Piccadilly went off forty minutes ahead of the others.

The second wave began shortly after seven. The gliders and their tows continued to leave at approximately five-minute intervals until 0230. Each glider was loaded to the absolute maximum weight they could take. It was a terrific strain on the C-47s as they circled on the long grind necessary to clear the mountains towing two gliders apiece, some of them carrying two and one-half tons apiece.

Before the operations were discontinued early Monday morning, fifty-two gliders had reached their destination delivering about three hundred fifty men and their equipment. Three crash landed shortly after take-off, three in the Imphal valley, and the remainder came down in Jap territory between the Chidwin and Irrawaddy rivers.

In every case, except the three that landed near the base, the gliders' tow lines broke. Although there was some very rough weather encountered along the route, the nature of the loose ends suggested sabotage of the lines.

The aerial reconnaissance of Broadway had failed to reveal many deep holes and logs in the high grass. As a result, there were a great many accidents when the gliders landed. Col. Alison and his crew quickly laid out a flare path in the smoothest part of the area, but two of the first gliders crashed on the improvised runway nevertheless. The interval of arrival was so short that it was impossible to clear away the wreckage of one glider before another would pile into the mess. Those who did see the wrecks of previous landings and swerved to avoid them frequently crashed into a ditch or on one of the scattered logs. One of the many freak accidents occurred when the pilot Brigadier Calvert's glider saw the wreckage ahead of him after he had already landed. He pulled back on the stick and hopped his ship over the heap, much as an adept golfer would overcome a stymie.

By shortly after midnight, the situation at Broadway had become so serious that Col. Alison radioed back a request that no more gliders be sent. The field was strewn with wreckage, and four men had been killed and ten injured in crashes in the jungle around the field, although this was not known at the time. None of the casualties was due to enemy action. Indeed, the enemy had not been encountered when I was at Broadway yesterday.

By early Monday morning, the operation appeared to have reached a very critical stage. Twelve gliders were down, three of the ten UC-64s which had flown in with food had failed to return, and there were two messages from Col. Alison and Brigadier Calvert, who sent the cryptic note "Sayalink," meaning "Bother on the ground." At nine o'clock, Col Cochran and General Wingate reestablished contact with Broadway and heard the message, "continue Thursday." We were relieved as if our team had made a touchdown.

During Monday, the fighters and bombers began systematic destruction of the Japanese Air Force facilities in the Katha area. These operations were successful, and no enemy aircraft were encountered.

On Monday night, a commando force headed by Lt.Col. Gaty went into "Chowringhee," the southern base. The twelve Air Commando Force C-47s made the trips with single tow gliders. Two Americans and three Britishers were lost in a crash in this operation when a bulldozer got loose on a crash landing. However, the base was successfully established, and a runway cut out by men of the 900th Airborne Engineers who went along. The first night of transport operations, March 7, one hundred and six transports were handled at the jungle base the Commandos had built.

During Tuesday, Lt.Col. Grant Mahoney, leader of the fighter section, and Lt.Col. Robert T. Smith carried out a reconnaissance flight to air establishments in the Mandalay area. They discovered that the Japanese had moved seventy or more planes into the area, and it was assumed that these were about to attempt to interrupt the American air operations.

The Commando B-25s had again attacked in the Kawlin area that day and had assisted in locating the wreckage of all the gliders down in enemy territory.

It was decided to strike at the Japanese Air Force as soon as possible, and a raid by fighters carrying two 500 pound bombs each was launched on Anisikan early Wednesday afternoon. In this attack, twenty-one P-51s destroyed or damaged between twenty-five and thirty enemy aircraft and seriously damaged the airdrome at Anisikan and Onbauk.

Immediately after the fighters returned, they began preparations for another trip as soon as the ships were armed and fueled. The medium bombers were prepared, also. The Commando headquarters had informed the Third Tactical Air Force of the Jap concentration, and it was decided that the latter's planes were to come into the Shewbo area when our planes had finished at eight thirty. It is an apt commentary on the completeness of the job done by Col. Cochran's men that the Tac Air Force informed Cochran next morning that the Hurricanes and Vengeances they had sent had been unable to carry out their attack because the heat and smoke of the fires left after the Commando attack were too intense.

The five Commando bombers had carried their racks filled with fragmentation clusters and alternated with incendiary clusters. They went over the Shewbo airdrome in line abreast and dropped their bombs. First fragmentation and then the incendiaries at 300 feet intervals from 1,000 feet altitude.

This flight was led by Lt.Col. R.T. Smith, who had been in the fighter raid a few hours before. The fighters and bombers were out over the same three Jap bases early the next morning and could not find anything on which to drop their bombs. Below them was a scorched area of wrecked planes; even the anti-aircraft fire was missing. Intelligence reports confirm thirty-four Jap planes shot down or destroyed on the ground by the fighters with two probable and two damaged, and twelve destroyed by the bombers with one probable and five damaged. Since this phase of First Air Commando operations, no Japanese aircraft have been found or reported in the Mandalay area or north of there to Mytikyina.

In the fighter and bomber operations in support of operation "Thursday," two fighters have been lost in combat and one B-25 had to make a crash landing at "Broadway" when an engine failed. The base established at "Chowringhee" was only temporary. It was therefore abandoned Thursday night, March 9, when the 111th Brigade which operated in that area was brought in.

"Broadway" is still being reinforced with RAF and U.S.A.A.F planes of the Troop Carrier Command and the transports of the Commandos. So far, the two organizations have flown eight thousand fully equipped men and from eight hundred to one thousand mules and horses to the two strongholds 175 miles into Japanese territory. One C-47 troop carrier has been lost due to a taxiing accident at "Broadway."

The total weight carried by the transports amounts to 1,668,800 pounds. Seventy gliders loaded to capacity have been towed in by the Commandos. All transport and glider operations were carried out by night, and the sick and injured have been evacuated by liaison plane or transport.

The intense initial phase of the Air Commando operations is very nearly complete. The ultimate contribution of the Commando operation to our success in the Southeast Asia Command is a matter of conjecture. But from a tactical point of view, what has been accomplished here in the past six days clearly demonstrates the value of the air task force. An entirely unique fighting organization that has been tested under the most trying conditions and has proven that it can operate effectively.

After Thoughts

The original cadre for the Air Commando Force were loaded with rank, and I do not recall many promotions given to the enlisted personnel. Almost all the officers did get promoted a grade, but the bomber crews who were brought in from other outfits did have crew members flying in the lower pay grades. An example was Pvt William J. Winn, who was killed in action with his entire crew when their plane crashed in Burma on March 4, 1944. It was unfortunate that none of the enlisted crew that were killed on this mission were listed in the roster of assigned personnel prepared by Group Headquarters, April 12, 1944. Tech Sgt William B. Postlewait, Staff Sgt Joseph B. Klaus, 1st Lt Murrell Dillard, Pilot, and 1st lt. Leslie Nielson, navigator, did make the roster.

At one point, our own crew chief was rather disgruntled because he was only a sergeant, so he decided to take matters into his own hands. One day, after fortifying himself with a bottle of India gin, he walked into Col Alison's office with the gin bottle in one hand and a machete in the other, and demanded a promotion. He didn't get the promotion, but neither did he get demoted. He continued on as our crew chief and never talked about this episode.

The enlisted men got rationed a case of beer at infrequent intervals, and the officers received a liquor ration. There was always a great deal of swapping going on between the two groups. We tried cooling down the warm beer in the hypo solution from the photo lab. This worked pretty well, but the solution left a residue on the cans which made for a terrible taste when drinking the beer from the can. I would stuff some cans in the breech of our .75mm cannon to cool them down during a flight. Later, an NCO Club was started, and we could turn in our beer for tokens to be exchanged for a cold beer.

We were all issued Escape and Evasion kits. Some of the items these kits contained were Mickey Mouse watches, dime store beads and silver rupees. I even saw a block of opium lying on the floor of our operations basha. It was wrapped in wax paper.

We were also allowed to pick up personal weapons from a collection of M1 rifles, .30 caliber carbines, Thompson sub-machine guns, and Brit hand grenades. I put a rifle and a Thompson in the radio compartment of our plane and also stashed a case of grenades in our dispersal area. (Jap ground forces had been making a drive into the Imphal area, and I didn't want to take the chance of getting caught short.)

Lt. Weber borrowed one of my grenades one evening before a poker game. Of course, I had defanged it. During the course of the hot poker game, he let it roll off the table. Everyone thought it was a live grenade. Although the people present got over their fright, he remained unpopular for sometime afterward.

Originally, we had received twelve B-25 aircraft. Of these, we lost five. One went down on a combat mission with the loss of the entire crew. One went down in a storm with the loss of the entire crew and passengers, including General Orde Wingate. They are interred in Arlington National Cemetery. One crash-landed and burned at Hailakandi, but the crew got out safely. One crash-landed at *Broadway* due to engine failure. This crew also escaped. One lost the nose gunnery hatch on a combat mission and damaged the fuselage when a bomb fell through the bomb bay door. A photographer was injured. The loss rate during the four-month operation was forty-one percent of our aircraft and sixteen percent of our air crews.

Chindits' Viewpoints

During a six-day stay at Bangalore, where we turned in our B-25s, most of the fellows talked to British ground forces who had been at Mawlu and the surrounding territory. They gave very good accounts of what our ships actually did in support of their operations.

One soldier said that on one raid we made at the Mawlu road block the Japanese had moved in a good many men, and that the Chindits were almost out of ammunition, and were on the verge of retreat, when the bombers and fighters came over. This raid carried a bomb load of twenty-two one hundred pound bombs in each B-25 and there were nine bombers, plus depth charges from the fighters. He said that we dropped our bombs directly into the midst of the Japanese, and he could actually see the bodies going up with the explosions. After we had bombed and they had sent patrols into the area, they counted over four hundred dead Japanese, 98% due to the bombing and fighter strafing. He also said that one soldier actually sat down and cried when he saw our bombers coming to their support.

Another time, the Japanese had a Chindit patrol cut off from their own lines. We enabled them to escape when we repeatedly strafed an area about two hundred yards wide by one thousand yards long. The Chindits ran through this area as soon as we stopped our runs, the Japs being either dead or in retreat.

Once while on a special mission at Indaw Lake we bombed two villages and then strafed an area supposedly containing trenches. Unknown to us, the Japanese had left the villages to escape the bombardment and taken cover in the trenches. As it turned out we knocked the hell out of them when we strafed those trenches and enabled the Chindits to gain access to a very important water hole.

The troops said our depth charge attacks were terrible on the Japanese. Sometimes they could get out of their fox holes and watch unless the Jap lines were too close. We unluckily shot and killed one British soldier during a strafing attack and wounded another in the leg. We also killed several of their mules.

More From my journal, March, 1944
Comments on Japanese Morale

There has been a definite drop in the morale of the Japanese soldier. The number of prisoners taken by our forces in this theater is relatively high for the local estimated casualties. Until this time nearly all P/Ws have been helpless because of wounds or sickness. In recent operations a noticeable proportion of unwounded men have surrendered. Further, the one common characteristic is the suprisingly low morale of all P/Ws. One P/W states that his battalion, including officers, was very low in spirits due to length of time in action without relief and lack of food and ammunition.

Another spoke of the great shortage of food and states that, once in action, the majority went without food, not knowing when they would next eat. He admitted that due to lack of food and rest their spirits were very low. One prisoner was told by his commanding officer he would be killed if captured. Still another man stated the morale of his regiment was bad. The older soldiers who had come to Burma at the beginning of the war had expected to go home in December 1943, but now found themselves involved in still harsher fighting and greater hardships. He stated there was much complaining from the troops because of shortage of ammunition and uncertainty of relief supplies. There appears to be plenty of ammunition in forward dumps, but the difficulty is getting it distributed to their troops. It is now very clear the lack of supplies and ammunition is one of the chief causes of poor morale.

Another important point is the severe criticism of officers not encountered before. One P/W stated that two of his commanding officers should commit *harakiri* to atone for their failures.

The following is a copy of Kubo Tai's order on Morale Training, Japanese O.C. 55 Division, 2 March, 1944:

"In recent operations every unit has listed a number of men missing. This is truly deplorable and therefore I issue the following order to be brought without fail to every man of Butais under my command:

1. For a Japanese soldier to be captured alive is a crowning disgrace. Every C.O. will warn his men that each man will swear to uphold his principles and not give capture a thought.

2. If, by any chance, a man, while unconscious, falls into enemy hands, no word must be spoken regarding our forces, his organization, armament, strength, or dispositions, and no action which could profit the enemy be undertaken, whatever punishment or kind treatment be received."

The Japanese have never issued such orders to troops before, and it shows a worry in the minds of their Staff Commanders.

The Eastern Air Command really tore the Japanese Air Force apart during the week of April. Our P-51s knocked out 25 Jap planes on Aungban airfield. It took just six minutes over the field to do this, plus destroying two anti-aircraft positions and starting numerous fires. Japanese casualties in the Imphal attack to date, 3lst March 1944, is two thousand, six hundred killed, excluding those killed by air action.

The B-25s and P-51s of the Air Commando Force have killed approximately four thousand, five hundred Japanese troops since start of operations. They have dropped approximately four hundred to five hundred tons of high explosive and fragmentation bombs. Reports from our ground forces also say that the Japanese troops are in deadly fear of us and have started burying their food supplies for protection. Their air force also shows a growing distaste in tangling with us, particularly our fighters. Our B-25s have yet to be attacked in over three months of continuous assaults on Burma targets.

12

Going Home

Soon after arriving at our new base, we had a formation for awarding the Distinguished Flying Cross and Air Medal. Unless you did something outstanding, the criteria in the 10th Air Force was fifty combat missions for the DFC and twenty-five for the Air Medal. Our crew was included, and after standing in ranks for a long period of time, the formation was dismissed. Our names were never called, and there were some embarrassed and angry flight crews.

Afterwards, we were notified that there had been some confusion with the orders. Two sets of orders had been sent out by Tenth Air Force Headquarters, and one set had not been received. There were sixty-eight officers and non-commissioned officers in the second set.

The next day one of the operations clerks called me and said, "Sgt. Baisden, your DFC and Air Medal are in. Don't forget to sign for them." The following winter, after going through B-29 gunnery school at Lowry Field, Colorado, I was formally awarded the Air Medal. I was never formally awarded the DFC.

Our aircraft, *Barbie* III, has been pictured in *Squadron Signal Publication No. 34*, and the crew were featured on the front cover of *The Cockpit*, an aviation-oriented catalog sales company. I have been asked a number of times why the tape was applied to the side of the aircraft, obscuring the name *Barbie*. The explanation is simple: the armor plating attached to the fuselage side was corroding into the fuselage skin due to the wet conditions and humidity of the area. The tape was applied to cut down the corrosion, but it was only a temporary solution. Soon after the pictures were taken, our crew chief painted over the tape, repainted the name *Barbie*, and I added the mission bomb symbols.

I had enough of India, Burma, and China to last a lifetime. I had been bombed and strafed. I had recurring backaches from my bout with malaria, and I had just

Chuck and Willa—Lowry Air Force Base, Denver, Colorado, October 1944. Studio Photo

completed what were some very hairy combat missions in a three-month period. I wanted to go back to the States, but I was ordered to the fighter section for a possible promotion to Warrant Officer just when the bomber crews were informed that bombers were not going to be used in further Air Commando operations.

When R.T. Smith left our group, he had requested that I be returned to the ZI (zone of interior, or U.S.A), but I was never told that. His request was evidently rejected, or someone just forgot about it. Since Colonels Cochran and Alison had left the group, I talked with our operations officer and mentioned that this was my second tour. He told me when the present operation was over I would be returned to the States.

Several days later, I was on orders going home with written recommendations for Warrant Officer approval. Flying into Casablanca, Morroco, I caught an Air Transport Command C-54. Just prior to boarding, I had to give up my issued paratroop jump jacket. It was not regulation Air Force issue. My journal was also confiscated, but was returned to me by the War Department several years later.

One of the Commandos returning with me had a folding stock carbine which he kept on a short sling and covered with a raincoat slung over his shoulder. How he got it into the States without being caught, I'll never know. The last I saw of him was on a street outside LaGuardia Air terminal, walking down the street with his raincoat and carbine still intact.

I left a .45 Colt automatic pistol and a fine pair of ten power binoculars on my bunk at Asonsol. We never signed for much of our equipment, but to me it wasn't worth the chance of getting into trouble to take it.

I received orders to report to the Atlantic City Redistribution Center in New Jersey. I had a twenty-day leave, so I took my bride of fourteen months and enjoyed a lazy August in the countryside of Pennsylvania. Then we both checked into the Ambassador Hotel, headquarters for the Distribution Center. I was processed through for new uniforms and a physical and records check. During one interview with the doctor who had treated me for malaria, it was decided that I needed more rest, so my wife and I spent twenty-one wonderful days at an Air Force rest center in Lake Lure, North Carolina. Upon our return to Atlantic City, we found ourselves in the middle of a hurricane, so I received another twenty-day convenience leave because things in Atlantic City were in bad shape.

Of course, I was asked for my preference of assignment, but I knew this really meant I was going to be sent where I was needed. I had requested an assignment to medium bombers or to a fighter unit. I could forget about my warrant officer promotion. They couldn't help. Finally, I got orders for B-29 gunnery school at Lowry, AFB, Colorado.

In early October, 1944, my wife and I arrived at Lowry in Denver. I was in charge of a draft of former B-17 and B-24 gunner returnees. The course here was supposed to take about three months, and then the gunners would be shipped to

Overseas Operational Readiness Squadrons for shipment to B-29 bomber units in the Pacific. It became immediately apparent that the gunners who had completed their combat tours were mad about doing a second tour when the "homesteaders" at Lowry had yet to do their first tour. For this reason, some of the gunners were deliberately failing the course.

A solution was found, however. If the gunnery students who had completed a combat tour finished the gunnery course and then successfully passed the turret gunnery maintenance course (an additional two or three months of schooling), they would be considered for assignment as instructors at Lowry. Several of us jumped at the chance, and we were made instructors after we completed the prescribed course. A number of the gunners who did not pass were shipped out to B-29 squadrons whether they passed the gunners course or not, along with some of the homesteaders.

Shortly before the war ended, my wife presented us with a baby girl we named Jerry Rae, after my favorite war correspondent's wife, whose name was Jerry. The correspondent was Ernie Pyle, who had recently been killed in the Pacific theater.

13

Civilian Life and Re-entry into Military Life and the Korean War

With some misgivings, but with the added responsibility of a new member in our family, and pressure from my father to come East and join him in his business, I joined with thousands of other veterans and was discharged on September 24, 1945. I received the veterans emblem promptly designated by all as "The Ruptured Duck."

We moved to a little village in the country outside of Scranton, Pennsylvania, where we lived in my grandfather's house that we had purchased while I was in the service. I worked as a watch repairman, taking advantage of the "on-the-job-training" that the government offered to veterans. This was a good program, but, unfortunately, many businesses did not keep their promises of wage advancement. The government "on-the-job-training" monthly payments decreased the longer you were training. At the end of two and one-half years, I was making less than when I started, and I was becoming quite disenchanted with civilian life.

So it was I went to Mitchel Air Force Base on Long Island and reenlisted as a Staff Sergeant. There was no chance of getting my Master Sergeant's rating back because so many former commissioned officers were being retained in the service in this pay grade.

We now had another child, a son, Dan, but even with the paltry pay and allowances of pay grade, I was better off in the service than I had been as a civilian. I never regretted this move, but I am very thankful I had a wife who stood by me. I knew it was very difficult for her.

In the fall of 1949, we moved from Mitchel to McGuire AFB, New Jersey. Soon afterward, our third child, Bill, was born. In the spring of 1950, I received orders for duty in Japan. My family could not travel with me. Taking my wife and three children and towing our thirty-three foot trailer, I headed for Nebraska, where

Lockheed F-80C jets from 80th Fighter-Bomber Squadron (HEAD HUNTERS), *armed with 5" Hi-Velocity Rockets (HIVARS) prepare for a mission from Itazuke AFB, Fukaoka, Japan, 1950.*

Author about to lose stomach in a simulated dive bomb attack in a Lockheed T-33 Trainer.

One of our F80s after tangling with a N. Korean cable.

Removing weapons and munitions from a taxi accident.

A F-80 that did not make it.

North Korean Yak fighter downed over Kimpo AFB, Korea.

my wife's parents now lived, and left them there. Then, I headed for my new assignment.

In early June of 1950, I took a troop ship to Yokohama, Japan, and to the 80th Fighter/Bomber squadron, 8th Fighter Group, located at Itazuke Air Force Base, Kyushu. Any hope of my family joining me were lost with the outbreak of the Korean War a couple of weeks after I arrived.

The Korean War

The 80th had Lockheed F-80C jet fighters which were very rugged aircraft, armed with six AN- M3 .50 caliber machine guns, zero rail rocket launchers for the five-inch HIVR (high velocity aerial rocket). This rocket was five inches in diameter and six feet long. It weighed one hundred-forty pounds with a warhead charge of about eight pounds. Its burning time was 1.2 seconds. The fin stabilized with a velocity of thirteen hundred feet per second and a maximum accurate range of one thousand yards. An instantaneous nose fuse could be installed with a fixed base delay fuse in the base of the warhead.

For flak suppression, we used a proximity nose fuse in the rocket heads and also in the five hundred and one thousand pound bombs. These fuses sent out a signal, and when the signal bounced back to the fuse head as it neared the ground, the charge was detonated. However, the rockets were prone to detonate when fired into thick cloud cover, also.

The K14 gun sight had to be set for either rocket or gunnery since the trajectories were not compatible. Should the sight be sighted for firing rockets and not reset for gunnery, the guns would fire over the target. Once in awhile, a hot shot pilot who had just returned from a mission would want to know what was wrong with his guns. It would prove embarrassing when I would reach over, reset his gun sight to the gunnery setting, and say, "Sir, you forgot to reset your gun sight."

We experienced a lot of trouble with rocket accuracy due to using rocket motors which had been in storage too long. As squadron armament inspector, I kept records of all armament malfunctions. At one time, we were getting one direct hit out of every fifteen rockets fired. Of course, a near miss was very effective against unarmored targets.

Four rockets could be loaded onto the launcher assemblies and four more attached to the rockets themselves, for a total of eight. A single one thousand pound bomb could be carried under each wing in place of drop tanks. A gun camera pod, using 16mm film, was installed on the right wing. During the first few weeks of the Korean War, we lost some of our aircraft due to the pilots trying to record their rocket hits on these cameras. They were caught in their own rocket bursts. The camera angles were later changed to correct this problem. We still had losses from ground fire, and from cables strung through narrow passes or defiles.

Engine from Russian Stormovik attack fighter. Very heavily armor plated. Kimp AFB, Korea.

Han River bridge at Seoul, Korea, destroyed by Air Force Bombers, 1951.

North Korean-Russian Type 33 tank destroyed just north of Seoul, 1951.

Church in Seoul destroyed by artillery fire, 1951.

Church World Service Buildings burnt out in Seoul, 1951.

Napalm, as used in drop tanks, was gasoline mixed with a powder that gave the gasoline the consistency of rubbery, sticky syrup. We used Japanese-made drop tanks and ignited the napalm with "Willie Pete," or white phosphorus grenades, replacing the regular spoon handle and delay fusing with an aneroid type spinner arming vane and a No.10 dynamite cap. Two were used, one in the filler cap receptacle and one on a bracket at the rear of the drop tank.

The 80th moved to Kimpo AFB (K14) just outside of Seoul, Korea, in the summer of 1950, and remained there until December, when the Chinese broke through at the Yalu. We evacuated the base and went back to Itazuke in time for Christmas, returning again to K14 early in the summer.

One time, after a flight of our F-80s returned from a mission, we were visited by the brass from Group Headquarters. They wanted an exact accounting of what weapons were used and how much ammo had been expended. No explanations were offered for this request, but rumor had it the fighters had gotten themselves lost and proceeded to strafe an airfield. Unfortunately, the airfield was in Vladivostok, U.S.S.R. The planes had been old American-made, lend-lease aircraft left over from World War II. This was strictly rumor, of course, but years later I heard that a court-martial had convened. The defendants, however, were dismissed because there was no one there to prefer charges.

I had been promoted to Technical Sergeant and had been promised the armament section when our present NCOIC was rotated back to the ZI, but it never happened. At this time a number of recalled Air Force Reserves arrived, including one Lieutenant (unrated) who was made our armament officer. Previously, all our armament officers had been pilots from our squadron. Unfortunately, our new armament officer knew very little about handling his men and less about armament. Then he made another of his buddies, a Tech Sergeant, who knew even less that he did, NCOIC of our armament section.

I was pretty fed up with this incompetence, so in the late summer of 1951, I had an opportunity to escape when a request for personnel with B-29 Central Fire Control experience came up. I put in for it, and was soon on my way to Okinawa to join the 93rd Bomb Squadron, 19th Bomb Wing. We were known to some as "General MacArthur's Private Air Force."

The request for gunners that had come to K14 was not exactly what the 19th Bomb Wing had wanted. They were short of complete crews, but they were receiving individuals that, for the most part, were not current in the B-29. This included pilots as well as gunners.

After a gunnery refresher course at Guam, I was assigned duty as a spare gunner. It didn't take long to learn that as a spare gunner, I was flying quite often. This was especially true during maximum effort missions when some gunners would suddenly go DNIF (duty not involving flying) with various ailments.

Author installing a proximity fuse in 500 lb. General purpose bomb. These were used against anti-aircraft installations.

Perimeter night guard duty at Kimpo AFB Dec 1950. Author found this type of duty non habit forming.

Income tax was exempt for each month that a combat mission was flown. Thus, we usually had several passengers (field grade types) flying with us on night missions. These night missions were considered "milk runs."

The weaponry in the Boeing B-29 bomber was, in my opinion, the best defensive system for bombers that came out of World War II. Although I did not have the affection for this aircraft that I had for the B-25, the bomber provided a defensive platform with a higher chance of survival for the crew. This was due to the accuracy of the computerized weapons, and because the gunners were not being exposed to frigid open compartments or cramped turrets. The aircraft was pressurized and could reach altitudes higher and fly faster than either the B-17 or B-24. They flew three hundred eighty miles per hour at twenty-five thousand feet and almost four hundred miles per hour at thirty thousand feet.

The General Electric fire control, or central fire control (CFC) system, was made up of five remote-controlled turrets, and three .50 caliber two-gun turrets: the lower aft, the lower forward, and the upper aft. A four-gun .50 caliber turret was located in the upper forward, and either two .50 caliber and one .20mm, or three .50 caliber machine guns, were located in the tail turret.

93rd Bomb Squadron-19th Bomb Wing—Kadena AFB, Okinawa, 1951.

Mileage post at Kadena AFB.

The Boeing B-29 Heavy Bomber from the 93rd Bomb Squadron.

19th Bomb Group bombing mission to North Korea, 1951.

Wing ship from author's position.

Gunners had more comfort in the B-29 than they did in the Boeing B17 and the Consolidated B24s.

Gunsight and controls for the right side scanner/gunner.

Bombing Wonson, North Korea, marshaling yards—12 September 1951.

Our bomb strike on Taechon Air Field, North Korea, 22 October 1951.

Three gunners were in compartments in the aft section of the fuselage, and on the right and left of the compartment were the scanner/gunners. The central fire control gunner was seated on a raised pedestal where he could observe from the aft top section of the fuselage. He was also in charge of the gunners. The tail gunner was isolated in a pressurized compartment of his own. The bombardier acted as gunner for the nose of the aircraft. Only the gun sights and necessary operating switches were in the gunners' compartments and the turrets were not accessible during pressurized flights. Control between the sighting stations and the turrets used the principle of Selsyn signals, the sight being the master and the turret the slave. This could be compared to a present day remote control television antenna. The sight itself was mounted on a yoke and stand, which allowed for elevation and azimuth movement. Horizontal ranging bars were presented in the sight image, and were controlled by a round hand knob on the side of the yoke body. The gunner kept his target framed, and he fired the weapons with a thumb switch. A gun camera was also mounted on the yoke assembly.

All movements of the sight were transmitted to a main computer, which solved the deflection angles and allowed for the proper leads. Gunners had only to track smoothly with the ranging bars kept on the wing tips or the outside of the fuselage of their target. There was an automatic gun charging system operated by a small air compressor. It would recognize a failure to fire and charge or recycle the gun mechanism within seconds. However, these had a tendency to freeze at high altitudes.

Control of the turrets used a primary, secondary, and tertiary method. Under normal flight conditions, the primary control would be the tail, gunner-tail turret, left side, gunner-lower aft turret, right side, gunner-lower forward turret, CFC, gunner-upper aft turret, bombardier, and gunner-upper forward turret. The bombardier could also take primary control of the lower forward for protection against frontal attacks. Each gunner was designated a search and fire area, depending on the position of the aircraft in the bomber formation. When a gunner pressed his spring-loaded action switch, a paddle-like disc mounted on the yoke, the turret he controlled would follow the movement sightings. Should he release the action switch, the turret would follow the movement of the gunner who had secondary control.

When the bombardier gunner released his action switch, the upper forward turret would follow the sight movements of the CFC gunner. The same thing would happen between right and left gunners. Tertiary gunner control was only from the tail turret to either the right or left gunners.

The B-29 was at its best during WWII, but during the Korean war we had problems with the Russian-built MIG jet fighter. By the time I rotated back to the States in December 1951, daylight bombing was becoming very unpopular. Unless we had good fighter cover, and I am only referring to the F-86, the MIGS worked us over.

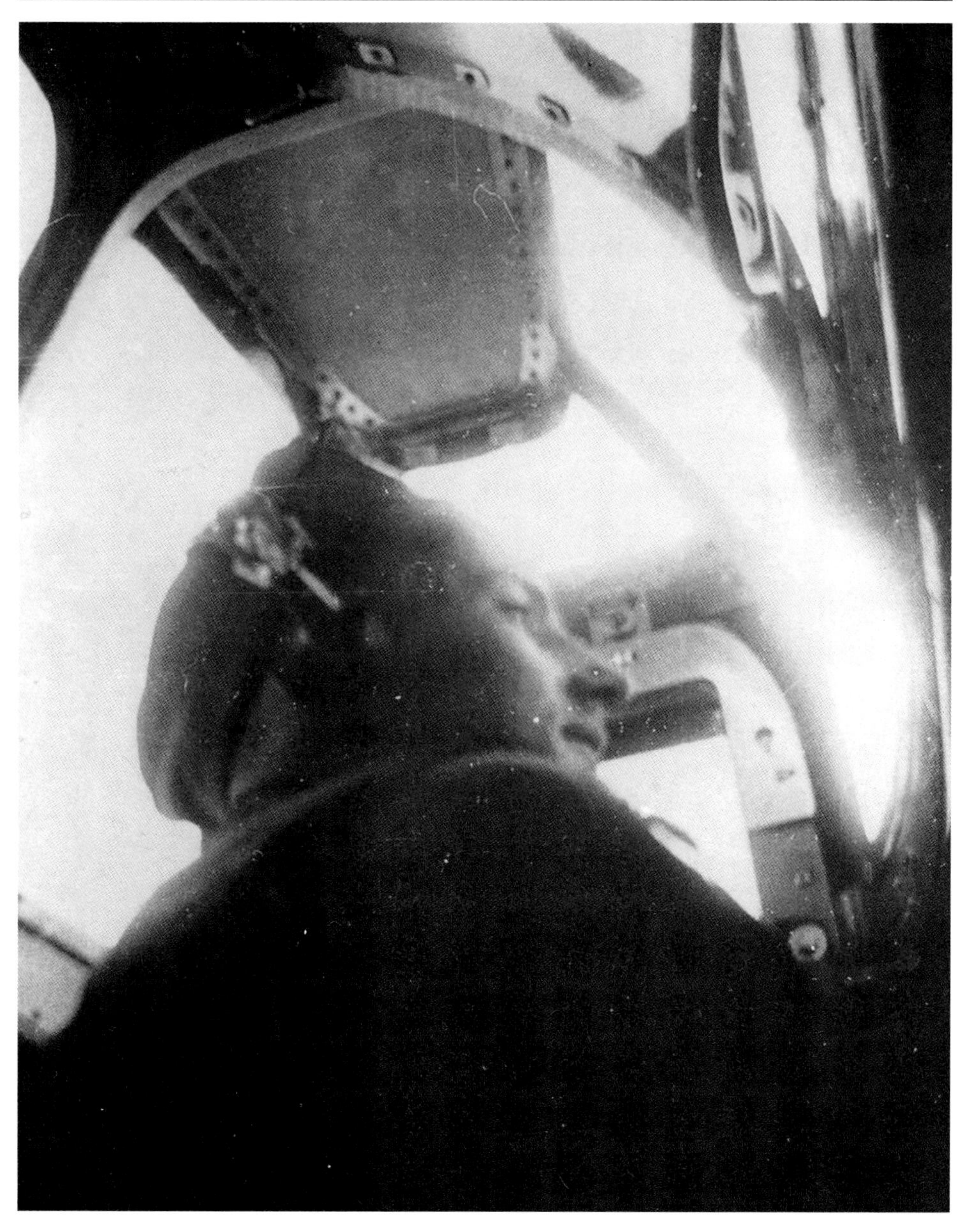

Tail gunner's home in the B-29. You could always see where you had been, but never where you were going.

While flying a daylight mission to bomb the Taechon Air Field in North Korea, our CFC gunner was credited with a MIG shoot down and received the DFC for his efforts. While I am admittedly not a military tactician, I did have some experience in the B-29 under combat conditions, and I have always felt it was not the best aircraft for that type of war. Most of our targets were too small for a large formation that gives the best defense against fighter attack, and the B-29 was too slow when compared to jet aircraft speeds. It got to the point where many gunners would fire on any aircraft that came in nose bearing on his aircraft. There was simply not enough time to figure out who the bad guys were. A MIG would come through the formation with his dive brakes opened so he could slow down enough to take a crack at you. Our radar jamming did give us some measure of safety against anti-aircraft fire.

I think back and remember how terribly alone I felt sitting back in the tail of a B-29 at 23,000 feet over North Korea in the middle of the night. I was cold, bored, sleepy, and somewhat apprehensive.

My chance to return to the ZI came when a CFC gunner from a crew that had completed their missions had requested to remain with the 93rd Squadron. Most of us lived for the day we would be rotated back to the ZI. This guy was living with a Japanese woman, and they were operating a cat house just off the base at Kadena. They were making a small fortune, and he was living like a king. This was OK with me.

I flew to Tachikawa, Japan, to bring a B-29 back to the States with us. On a test hop, two engines overheated, and we found metal particles in the engine oil sumps. Our aircraft commander refused to take the aircraft or wait for the engine changes, so we caught a Northwestern chartered C-54 back to San Francisco.

14

Strategic Air Command (SAC) and the Cold War

I had been assigned permanent change of station (PCS) to the reactivated 308th Bomb Wing at Savannah, Georgia, with a ninety-day temporary duty (TDY) at Forbes AFB, Kansas. I arrived at Hunter in the spring of 1952. The 308th was a former bomb wing in the 14th AF under General Chennault in China and had been deactivated at the end of WWII.

The Wing made one forty-five day TDY to Morroco, Africa. We lost one B-29 in the Azores when it crashed into a mountain making a wrong turn in a traffic pattern in bad weather. My crew flew Wing Standardization Crew until the B-29s were phased out in 1953. I flew my last B-29 flight on September 4, 1953.

Gunners had the choice of early out on their enlistment assignments to a B-52 Bomber Squadron as tail gunners or taking on-the-job training as boom operators (In-flight refueling technicians). I chose the latter, and in due course, I was checked out in the KC-97 tankers. It turned out to be a wise choice, because the gunnery field promotions in the tech and master sergeant ratings were few and far between. Some of the gunners had been staff and tech sergeants for many years. When we finished training, many former gunners made master sergeant soon afterward. Twelve years after making master sergeant the first time, I again had six stripes and a job I really liked.

In-flight refueling using Boeing's flying boom system came into the Air Force inventory in 1949 with the Boeing KB-29P. In 1957 the Boeing KC-97 tankers replaced the 29Ps. In the KC a crew of seven was comprised of the pilot and aircraft commander, co-pilot, navigator, flight engineer, radio operator, in-flight refueling technician or boom operator (boomer), and assistant boom operator.

The boom operator's refueling position was on a padded, horizontal tray in the rear of the tanker. With his right hand he could move a control stick for elevation

A gaggle of KC-97F refueling tankers on a mass refueling effort.

KC-97F tankers from the 308th Air Refueling Squadron—Hunter AFB, Savannah, GA, 1954.

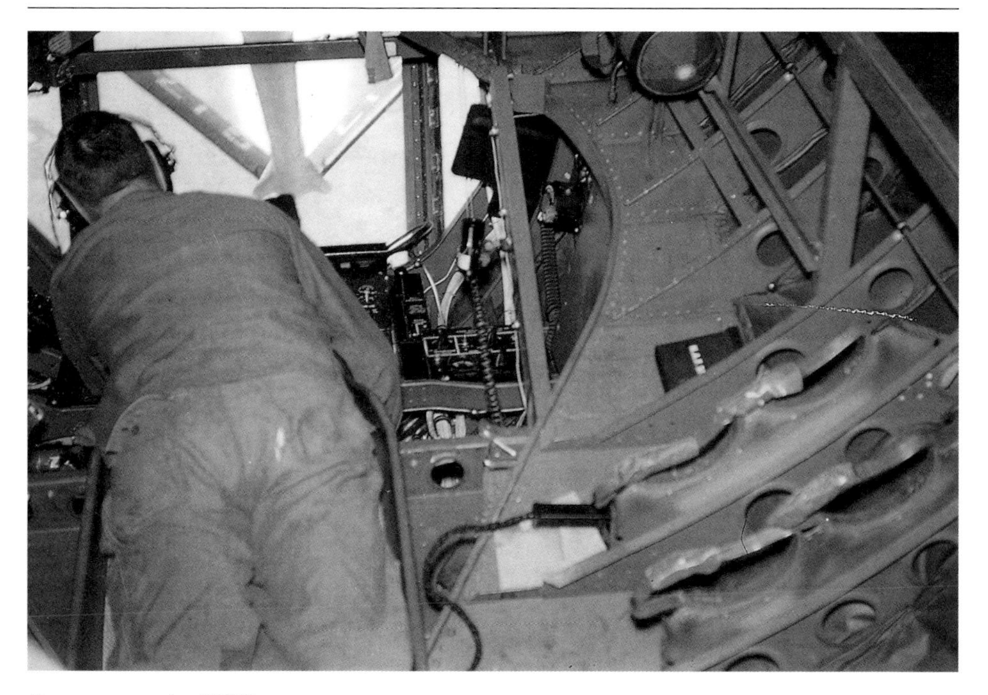

Boomer at work—KC97.

Flight engineer at work—KC97

Sometimes we would dump a little fuel to help the bomber crew locate us.

Boeing B-47 bomber closes in for refueling.

Contact tanker-fuel pressure coming up.

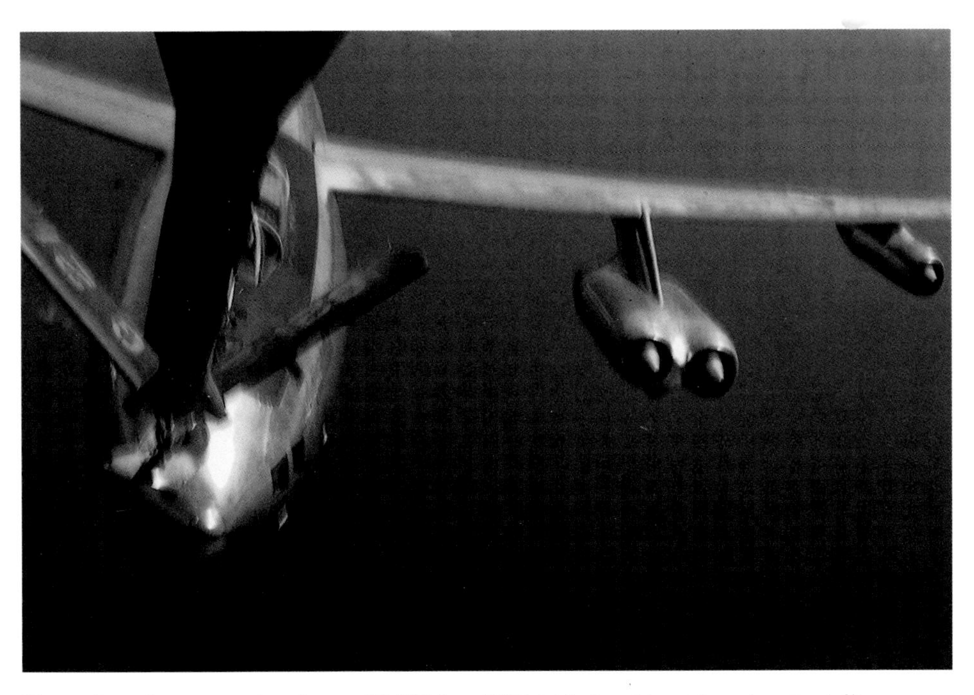

The reciever has now recieved some 35,000 lbs of JP4 jet fuel and is getting close to stalling.

Good bye and good luck.

and lateral boom movement. His left hand controlled a lever for boom extension. A trigger switch on the handle of the joy or control stick allowed for disconnecting the boom nozzle from the receiver aircraft. There were also instrument gauges for reading boom movements. A lighted panel under the fuselage gave the receiver pilot instructions when in the boom envelope. Fuel was transferred to the receiver by the flight engineer. Our average refueling altitude was between sixteen thousand and thirteen thousand feet, and we could get off thirty-five thousand pounds of JP-4 jet fuel in fifteen minutes when everybody had their act together.

We refueled B-47 and B-52 SAC bombers. It was not SAC's policy to refuel TAC fighters. But on one occasion I refueled an F-84 that was having fuel problems. We also towed this aircraft, on the end of the boom, back to his field at Warner Robins AFB, Georgia. He was one happy fly-boy.

Once the receiver was in the boom envelope and contact was made, the real work of refueling fell on the bomber pilot. It took real concentration and a lot of skill to complete a refueling without several disconnects, which was automatic should the receiver exceed movement limits.

Although the KC-97 tanker was a work horse, its disadvantage in refueling jet bombers was the prop driven tanker could not get to the bomber's best fuel conserving altitude, and it was too slow for jet aircraft speeds. As fuel was transferred the weight of the bomber increased, and at max tanker speed, the bomber would begin to stall out, and often did. Many times before completing an off load our tanker would have to go into a descent to increase our air speed and to prevent the bomber from stalling. A good boomer could really help a bomber pilot with clear, calm directions and a few words of encouragement. However, many times we refueled in complete radio silence, both in daylight and at night. The coming of the KC-135 eliminated these problems and also brought the demise of the KC-97 as a tanker in SAC.

The advent of the Cold War and General Curtis LeMay brought about the concept of twenty-four hour alert readiness. Alert facilities at all SAC bases were constructed. They were known as the "mole holes," where complete combat-ready crews would eat, sleep, and live together. The crews would rotate every week. The period of time to be airborne was so critical that minimum-interval take-off (MITO) was established. An aircraft would start his take-off roll before the aircraft ahead left the runway. Forty-five and ninety day TDY assignments were quite common to such places as Bermuda, the Azores, Greenland, Morroco, and Newfoundland. These TDYs also caused the divorce rate in SAC to escalate.

General LeMay also established crew and support professionalism by periodic testing, crew survival training, and a non-commissioned officer academy.

The 308th Air Refueling Squadron was disbanded in 1960, and in March of that year, I was transferred to the 19th Air Refueling Squadron at Otis AFB in Massachusetts. I had been there four years when I retired in 1964. I did not mind the flying part, but the constant and repeated duty of being on alert status which required living in the "mole hole" and the constant TDYs took its toll on me and my young, growing family. With my twenty years completed, I decided to leave the Air Force for good. On January 7, 1964, I made my last crew flight.

From 1954 to 1964 I had off-loaded some one thousand-eight hundred and twenty-five tons of JP-4 jet fuel and had made eight-hundred and fifteen refueling contacts with SAC aircraft. My Air Force career ended with one hundred and five hours of pilot time, two hundred and ninety-nine hours of combat time, and some two thousand air hours in KC-97 tankers. This broke no records by any means, but I was satisfied.

I retired with no regrets of service to my country and no regrets on leaving the Air Force, except to miss the camaraderie of the fine professional airmen I had known.

KC-97 Crew In 308th ARS Receives SAC Crew-of-the-Month Award

HUNTER HERALD

l. II 23 June 1954 · No. 35

AC's April-Crew-Of-The-Month . . . .

eft to right) Capt. James A. Nelson, aircraft commander, 2nd Lt. Donald A. Bonkessel, Jr., co-lot, 2nd Lt. Richard W. Johnson, navigator, TSG James Q. Pence, flight-engineer, A/2c George Floyd, Jr., radio-operator, TSG Charles N. Baisden, boom-operator and A/1c Marvin J. Gerber, sistant boom-operator. All are members of the 308th Air Refueling Sq. (Official USAF Photo by G Gerald Hickens)

ELSON-Retired as Lt/Colonel-deceased 1991
ORNKESSEl-Retired as Colonel.
OHNSON-Believed killed in Air/Sea Rescue plane crash.
ENCE-Retired as Chief Master Sgt.
LOYD & GERBER-Discharged at end of enlistments.
aisden-Retired as Master Sgt.
ositions of radio and assistant B.O. were later eliminated.
oomers were also Load Masters and trained in the use of the
extant to aid Navs. with celestial position fixes.

Capt. Nelson And Crew Land KC-97 After A Series Of Emergencies

HQ SAC (SAC PS)—Landing with no attendant damage to their aircraft, after a series of inflight emergencies during which the crew had been alerted to bail out, earned SAC's April Crew-of-the-Month award for a KC-97 crew commanded by Capt. James A. Nelson, 308th Air Refueling Sq., it was announced in the May Combat Crew Magazine.

Following a normal night take-off on a directed air refueling mission, the engineer, TSG James O. Pence, reported a prop warning light on the plane's number three engine. The tanker was in a 270 degree climbing turn and Captain Nelson directed the engineer to replenish as the climbing turn was continued.

Then the engine ran away to [illegible]00 RPM and when the co-pilot, 2nd Lt. Donald A. Bon Kessel, Jr., was unable to hold in the feathering button, the rate of climb was increased to slow the aircraft. At this time a decision was made to freeze the engine.

The co-pilot called the tower, declaring an emergency, and a slow turn was started back to the field. An inoperative dump valve made it impossible to relieve the critical gross weight, and the tanker was dropping 200 to 300 feet per minute. At the suggestion of the engineer, an attempt was made to remove the vacuum relief valve in an effort to get the fuel off-loaded.

Due to extreme buffeting this was impossible and at an altitude of 1,600 feet the crew was alerted to bail out. At this point the field was sighted and Captain Nelson elected to try for a landing downwind.

At 800 feet as the gear was being extended, the engine finally froze and stopped windmilling, Gradual letdown was continued and touchdown made approximately 2,000 feet down the runway. A normal landing and roll were completed.

Subsequent inspection disclosed that the oil return line coupling of No. 3 propellor had come loose.

Other members of the crew are 2nd Lt. Richard W. Johnson, navigator, A/2c George J. Floyd, Jr., radio operator, TSG Charles N. Baisden, in-flight refueling operation and his assistant A/1c Marvin J. Gerber,

SAC Crew of the month April 1954.

The Story Behind the Photo

One of the preflight checks prior to a refueling flight was to ascertain that the fuel dump mechanism on the boom was operational. A visual check showed that the fuel dump actuator (finger) was missing. This meant that there was no way we could get rid of some thirty-five thousand pounds of JP-4 off load fuel in case of an emergency.

I reported these findings to the flight engineer, who in turn reported it to the Aircraft Commander, who reported this to the Squadron Operations Officer, who reported this to the Wing Director of Operations. The word came back verbally to take the aircraft anyway.

The Aircraft Commander has the final say as to taking any aircraft, but this was the Strategic Air Command, and any mission that did not go as briefed was carefully examined in an unfriendly way by those higher up in the pecking order. In this case, the DOO was a full Colonel with the nickname "Bull," although we never called him this to his face. (He ate Captains for breakfast.)

There was always the priority of the mission as compared to flying safety, but in a peacetime situation, flying safety rated foremost. However, calculated risks were sometimes taken because of fear, namely, the fear of the wrath of the chief of the pecking order.

It was only the skill of the skipper and a lot of luck that prevented the aircraft from being destroyed. Our flight lasted just thirty minutes from our takeoff to touchdown. We did not even blow out a tire.

We lucked out and made SAC crew of the month.

Epilogue

Looking back, I realize now that I was part of a covert operation sponsored by my government. If this operation had been made public, it might have led to the impeachment of the President.

I am proud to have been a member of the American Volunteer Group (AVG), but at the same time, I do not believe this type of operation is advisable. If the Japanese had not attacked Pearl Harbor and America had not found itself at war with Japan, the outcome for the AVG would be, in my opinion, debatable.

It took fifty years for our government to admit the AVG was actually a covert operation with approval straight from the President of the United States. Former AVG members at their own request may apply for a Honorable Discharge from the United States Air Force for the period of December 7, 1941, until July 18, 1942.

In 1988 the Flying Tiger Association granted the use of their name and logo to the 229th Helicopter Attack Regiment, United States Army. This unit is a highly disciplined, well trained regiment, with an "Esprit de Corps" that is enviable. The Air Force Special Operations of today are the Air Commandos of yesteryear. They are the "quiet professionals" and tops in their trade.

In August of 1992, the Flying Tigers of the American Volunteer Group were presented with the Presidential Unit Citation and the Air Force Association Award. On December 7, 1996, all ground crew members of the AVG were awarded the bronze star, and all pilots, The Distinguished Flying Cross. These awards were presented in special ceremonies in Dallas, Texas, by General Ronald R. Fogleman, U.S.AF (Ret.), former Chief of Staff of the United States Air Force.

Author recieving the Bronze Star from General Ronald R. Fogleman, Chief of Staff USAF, December 7, 1996.

References

75 *MM Gun Training Manual* 9-312, War Department, 1944.

Air Force Manual 136-6, U.S. Government Printing Office, 1949.

Baisden, Charles. *Personal Journal,* Form 5 Flight Logs, Official Orders, Air Force and American Volunteer Group.

Military Field Branch, Military Archives Division, Washington, D.C.

Office of Air Force History. *The Army Air Force in World War Ii Combat Chronology* 1941-1945, Office of Air Force History Headquarters, U.S.AAF 19734, Air Commando Unit 5-3-4.

Pistole, Larry. *Pictorial History of The Flying Tigers,* Moss Publications, Orange, VA.

Smith, R.T. *Tale of a Tiger,* Tiger Originals, 14756 Leadwell Street, Van Nuys, CA.

Squadron Signal Publications No. 34, *B-25 Mitchel in Action.*

The Army Air Force Historical Foundation-Historical Times, Inc. Impact. *The Army Air Force Confidential Picture History,* Harrisburg, PA17105.

VanWagner, R.D. *Military History Series* 86-1, 1st Air Commando Group, Air Command and Staff College, Maxwell AFB, Montgomery, AL 36112.

Wagner, Ray. *American Combat Planes,* Doubleday.

Other Books of Interest:

Bedford, Timothy B. *The World War* II *Fact and Quiz Book,* Harper & Row, NY.
Bond & Anderson. *A Flying Tigers Diary,* Texas A & M University Press.
Chennault, Anna. *Chennault and The Flying Tigers,* P. Erikison, NY.
Hotz, R.B. *With General Chennault,* Coward, McCann, NY, 1943.
Howard, James. *Roar of a Tiger,* Orion Books, NY.
Losonsky, Frank S. & Losonsky, Terry M. *Flying Tiger, A Crew Chief's Story,* Schiffer Publishing Ltd. 77 Lower Valley Rd. Atglen, PA 19310.
Rosbert, J.R. *Flying Tiger Joe's Adventure Story Cookbook,* Giant Popular Press, Franklin, NC.
Shilling, Erik. *Destiny,* ISBN 1-882463-02-1.
Smith, Robert M. *With Chennault in China,* Schiffer Publishing Ltd. 77 Lower Valley Rd. Atglen, PA 19310.
Szuscikiewicz, Paul. *Flying Tigers,* Gallery Books, 112 Madison Ave. NY, 1990.
Wheland, Russell. *The Flying Tigers,* Viking Press, NY. 1943.

Videos:

Fei Hu Films, *Fei Hu (Flying Tigers),* A 90-minute video, 1993, 735 State St. Suite 631, Santa Barbara, CA 93101.

Also from the publisher

WITH CHENNAULT IN CHINA

A Flying Tiger's Diary

Robert M. Smith

Here's the story of how a handful of young Americans, fighting with improvised equipment, commanded the air against superior enemy forces and won! Written by a radio operator who served as a member of the AVG (American Volunteer Group) throughout their existence, this fascinating, intimate story of General Claire Lee Chennault's "Flying Tigers" is loaded with original photographs and numerous first-hand accounts from the author's personal diary.

Size: 6" x 9", over 110 b/w photographs
176 pages, hard cover
ISBN: 0-7643-0287-6 $29.95

FLYING TIGER: A CREW CHIEF'S STORY
- THE WAR DIARY OF AN AVG CREW CHIEF

Frank S. Losonsky & Terry M. Losonsky.

This book is the war diary of a Flying Tiger American Volunteer Group crew chief from the 3rd Pursuit Squadron. *Flying Tiger* will give aviation historians new insights into the days shortly before the Flying Tiger successes in late 1941.

Size: 8 1/2" x 11", 200 photographs
112 pages, hard cover
ISBN: 0-7643-0045-8 $35.00